a LIVING | FREE guide

Vertical Vegetable Gardening

by Chris McLaughlin

ALPHA
A member of Penguin Group (USA) Inc.

For my fellow gardeners who, like me, love the feel of the earth in their hands.

"Make voyages. Attempt them. There's nothing else." —Tennessee Williams

ALPHA BOOKS

Published by Penguin Group (USA) Inc.

Penguin Group (USA) Inc., 375 Hudson Street, New York, New York 10014, USA • Penguin Group (Canada), 90 Eglinton Avenue East, Suite 700, Toronto, Ontario M4P 2Y3, Canada (a division of Pearson Penguin Canada Inc.) • Penguin Books Ltd., 80 Strand, London WC2R 0RL, England • Penguin Ireland, 25 St. Stephen's Green, Dublin 2, Ireland (a division of Penguin Books Ltd.) • Penguin Group (Australia), 250 Camberwell Road, Camberwell, Victoria 3124, Australia (a division of Pearson Australia Group Pty. Ltd.) • Penguin Books India Pvt. Ltd., 11 Community Centre, Panchsheel Park, New Delhi—110 017, India • Penguin Group (NZ), 67 Apollo Drive, Rosedale, North Shore, Auckland 1311, New Zealand (a division of Pearson New Zealand Ltd.) • Penguin Books (South Africa) (Pty.) Ltd., 24 Sturdee Avenue, Rosebank, Johannesburg 2196, South Africa • Penguin Books Ltd., Registered Offices: 80 Strand, London WC2R 0RL, England

International Standard Book Number: 978-1-61564-183-3
Library of Congress Catalog Card Number: 2012944493

15 14 13 8 7 6 5 4 3 2 1

Interpretation of the printing code: The rightmost number of the first series of numbers is the year of the book's printing; the rightmost number of the second series of numbers is the number of the book's printing. For example, a printing code of 13-1 shows that the first printing occurred in 2013.

Printed in the United States of America

Publisher: *Mike Sanders*
Executive Managing Editor: *Billy Fields*
Senior Acquisitions Editor: *Tom Stevens*
Development Editor: *Lynn Northrup*
Senior Production Editor: *Janette Lynn*
Copy Editor: *Cate Schwenk*
Cover/Book Designer: *Rebecca Batchelor*
Indexer: *Johnna VanHoose Dinse*
Layout: *Brian Massey, Ayanna Lacey*
Proofreader: *John Etchison*

Cover Images:
Fresh Peas in the Pod © Masterfile
Cherry Tomatoes © Masterfile
Spice rack and Bamboo A-Frame © Chris McLaughlin

Contents

Appendixes

Introduction

I've always gardened in places where there were more people than cows or tractors. Garden beds in these places were never as big as I'd planned and were soon devoured by just a few plants. So they filled themselves up and I tried to do things differently.

I wish I could say that I had a stroke of genius or like some crazed, brilliant, horticultural Einstein I began sketching structural masterpieces on the back of napkins. No, the truth is that I'm extremely average. I learned about plants, soil, and microclimates by trying, failing, and finally succeeding. I still fail, and I still succeed—gardening is cool like that.

I gleaned the lion's share of any information I have from the generous gardeners I've met along the way. The clever ideas in this book are no exception. We gardeners are zealous in our efforts to share seeds, cuttings, stories, recipes, and advice.

After all, the *whole point* of gardening is to grow things, and a gardener's personal growth is perhaps the most important. The only way I know for that to happen is to question, attempt, discover, and share. Gardeners are excellent at these things.

So come on in and see what the creative gardening folk all over this country have come up with so that everyone can plant, grow, and harvest crops in their own backyard no matter where they reside.

How This Book Is Organized

Vertical Vegetable Gardening is divided into four parts:

Part 1, The Beauty and Bounty of Vertical Gardening, explains why vertical vegetable gardening is one of the smartest gardening techniques around, and what makes it so versatile. Less work, less weeding, fewer pests, less time, less money, and more produce is what a vertical vegetable garden has to offer you. In this part I explore the advantages to gardening vertically, discuss how plants climb and what structures work well, and give advice for where to locate your garden. I also tell you how to frame a simple garden bed and create other structures from scratch, and discuss creative ways to repurpose items you may already have around your house.

Part 2, The Basics: Soil and Seed, explores the fundamentals of gardening: defining soil, compost, and amendments. I include directions for traditional composting as well as composting directly in a bed (the compost sandwich). I explain warm- and cool-season vegetables and which seeds can be planted directly into the garden bed and which you may want to start indoors first. I also discuss how to start seeds and take cuttings so that you may never have to buy another plant. If you're curious about the differences between heirloom, open-pollinated, and hybrid seeds, I've got that covered in this section, too.

Part 3, Tending the Vertical Vegetable Garden, is all about the day-to-day workings of your vertical vegetable garden. Feeding and watering are clearly important for successful gardening, but there are a lot of questions that can come with it. Part 3 answers these questions, with details on the essentials of plant and soil nurturing. You find information on mulching, organic amendments, crop rotation, and pruning. I also explain organic and least-toxic pest control, and if you have to break out the big guns, I tell you how to handle those chemical products safely.

Part 4, Vegetables and Fruit That Enjoy Growing Up, is dedicated to the vegetables, herbs, fruits, and berries that thrive in a vertical setting. Each plant profile discusses planting, tending, harvesting, and the best-bet varieties, so you can choose what works best for you.

Extras

Throughout the book, you'll find the following sidebars that highlight information I want to be sure you'll catch:

The longer, individually titled sidebars add further thought on a topic within that section. They're often instructional in nature.

The **Good to Know** sidebars offer little pieces of advice and some extra food for thought on various topics.

The **Downer** sidebars alert you to some of the pitfalls you might encounter so you're not caught with your plants down.

Throughout this book you'll also find sections that are titled **Best Bets**. Those sections discuss which techniques or plant varieties seem to work extremely well in a particular situation. Best Bets are based on what is working for a large group of seasoned gardeners. That's not to say that it's the "end all" or "the answers of answers."

Best Bets are meant to guide you to either a situation or plant that many gardeners find superior. That said, I hope that you're like me and enjoy letting the experimenter inside you come out to play once in a while. If you have an idea or an instinct about something, I encourage you to *go for it,* and if it works well, to share your discovery with other gardeners.

Acknowledgments

No matter how many times I'm in the middle of a book, it always feels like the first time. I want to thank my extremely patient editor, Tom Stevens, for reminding me that it isn't (the first time). Thank you to my agent, Marilyn Allen, for her guidance and friendship.

A big shout-out to Alpha Books for taking on another fabulous book series; *Living Free* is a topic that's near and dear to my heart. Thanks to the rest of the editing team, Lynn Northrup, Janette Lynn, and Cate Schwenk, who are my second, third, and fourth set of eyes.

Huge gratitude to my daughter Hollis for actually keeping up with my line art requests, and to my family for putting up with the quirkiness of my job. Two high fives to the DGS who are by my side day and night (mostly night)—you know who you are.

I'm indebted (literally) to Jennifer Hammer, Shawna Coronado, Theresa Loe, Anthony Deffina, Jacky Alsina, Glenda Mills, Katie Elzer-Peters, Jayme Jenkins, and Annie Haven for their generosity in sharing their personal photos and structure contributions.

Part 1

The Beauty and Bounty of Vertical Gardening

Less work, less weeding, fewer pests, less time, less money, and more produce is what vertical vegetable gardening has to offer you. There are so many benefits to gardening vertically—it'll make you wonder why you would do it any other way. Discover why this is one of the smartest gardening techniques and what makes it so versatile.

Part 1 gives you the basics, helping you determine the best place to locate your garden and which containers and plants are most suited to vertical gardening. You'll even learn how to build your own structure from scratch. You don't need to spend a lot of money to get started; I encourage you to "shop" at home to look for items you can repurpose. I show you many ways to take those discarded items and use them in a new, creative way to help get your vertical garden off the ground.

The Vertical Advantage

1

In recent years we've all become aware of the advantages of growing fresh food in our own backyards. We've been given undisputable proof that fresh, homegrown food is tastier, healthier, and cheaper than purchasing it from a local grocery store.

It's easy to see why gardeners are touting the brilliance of getting back to basics and growing at least some of their own produce. But for those with what seems like precious little space, it feels like an unrealistic endeavor. I mean, how much food can you really get from an urban 4' × 4' spot of earth? Or a half barrel?

The technical answer will vary, because you have to take into consideration the plant species, variety, and climate. Yet, the honest-to-goodness-answer is: plenty. No matter how much space there is, your garden can produce plenty of food (even if only supplemental) for you and your family if you're willing to let your mind wander *up* instead of *out*.

The Perfect Produce Practice

If you're interested in gardening on any level, it's nearly impossible not to notice that vegetables are once again enjoying the gardening spotlight. In fact, food gardening hasn't been this popular with the general public since the Victory gardens that surrounded World War II. Even guerilla gardeners who once practiced random acts of beauty by planting flowers and shrubs in vacant city lots have now added vegetables, herbs, and fruit tree grafts to their illicit repertoire.

Vegetable gardening is good for you both physically and mentally, plus it offers optimal control over what you feed yourself and your family. Still, planting, growing, and harvesting food in what we think of as the traditional way (acreage and long hours in the sun), has been pushed aside for more logical gardening practices that work a little better for today's gardener or farmer. Large expanses of cultivated land have been swapped out for raised beds, containers, and one of the easiest and most rewarding veggie gardening practices—vertical gardening.

Naturally, the first reason that gardening vertically makes sense is that many of us are quite limited as far as gardening space. We have what's commonly referred to as small-space gardens and would like to get the most out of every square inch of soil. In fact, when I lived in the suburbs, that's exactly why I started gardening vertically: I had limited places around my home to plant a vegetable garden and wanted a decent-sized harvest.

Today, I live on 5 acres and I can honestly tell you that I grow more vegetables vertically than I ever did when I had much less land. The truth is that the space above the soil is underused as a growing resource, and it offers some surprising benefits to the gardener. (You just have to stop looking at the horizon and start thinking vertically.)

You're going to love the advantages of growing vegetables vertically.
(Photo courtesy of Katie Elzer-Peters)

Less Really Is More

It's obvious that growing vegetables up instead of out saves space. It opens gardening doors for people living in apartments and condominiums, as well as anyone with limited yard space (including those of us living in urban cities and suburban towns).

Lack of space is certainly a great reason to start thinking vertically, and that may be the road that led you to consider growing things vertically. But, it isn't the only good reason by a long shot. You may have plenty of room to plant a garden, but you would like to take advantage of any number of compelling reasons to add vertical components.

Less Time and Work

This reason alone is enough to keep my interest. Between raising kids, working, keeping house (and yard), paying bills, cooking meals, caring for pets, and volunteering, I lead a very full life, and my guess is that you do, too. Gardening and growing fresh food is something that I strongly believe in and have no intentions of cutting out.

The question is, how many things can I grow? Most of us keep up a busy daily pace just to stay afloat, so it may feel like one potted pepper plant is all you can manage. This is the beauty of growing vertically—the time commitment is very little compared to what's considered a horizontal garden bed. Of course, how much time depends on how many vertical gardens you're tending.

I should point out that even if you choose to have a large garden of vertical veggies, you'll still get twice as much done for your vertical plants as you would their horizontally grown counterparts. This is because there's very little soil for you to deal with, especially if your veggies are in a container. Less soil means less time watering for those of you who are hand-watering. Pruning plants such as berry canes, tomato plants, or fruit trees is easier. And harvesting? Harvesting is a quick endeavor when fruit is at eye level and can be easily seen and picked.

In short, your back and knees will thank you for adopting an upward gardening plan! Each of these factors also make vertical gardening the perfect method for those with physical limitations, as well. Gardeners in wheelchairs or with other physical challenges find that growing veggies up makes their hobby much easier—or perhaps even possible.

Less Money

Personally, this is a deal-maker for me. It's a tough economy, right? If you intend to create raised garden beds, growing plants vertically will save you money on purchasing soil because you won't need to build large rectangular beds. In fact, you'll be able to get away with obtaining just enough soil for the roots of the plants. When you garden with large, horizontal ones, you're providing fresh soil for the vines that simply rest on the soil as they sprawl; soil that's basically wasted.

The same principle applies to compost. Compost is the best thing you can do for your garden and whether you have your own compost piles going or plan to purchase this important amendment, it'll go a lot farther when you're adding it only to the area that really needs it—the plant roots.

I believe that no single thing benefits plants more than rich, crumbly compost. In Chapter 6, I tell you much more about compost and explain how to make your own so that you always have it on hand.

As far as building material for the upright structures that your plants will climb—as well as containers for plants to grow down—this may be the area where most of your dollars go. However, this isn't necessarily so. In Chapter 5, I give you plenty of ideas to push your imagination into overdrive on ways to recycle and upcycle items for the vertical garden that otherwise would have been discarded.

GOOD TO KNOW

Smaller beds mean a smaller growing surface. Combine this with a root delivery water system (such as a drip system) and rich, loamy soil to retain moisture, and you'll have a smaller water bill, too.

Fewer Weeds, Pests, and Diseases

One of the best perks of vertical gardening is that you'll have very few weeds sprouting up and even when they do rear their ugly heads, they can all be yanked out in minutes. With horizontal beds you're also weeding all of the bare soil areas in-between the plants so that they don't take over the garden as they mature. With vertical gardening you're working with much less soil surface, and many times you're starting with bagged soils that are weed-free from the outset.

Plants grown vertically up (such as trellising) or vertically down (such as hanging baskets) enjoy exceptional air circulation, much more so than their ground-dwelling counterparts. More air circulation around the plant foliage means less trouble with pests and disease, which means a stronger plant that will produce more unblemished fruit. And much, much less food waste due to rotting.

When plants are grown horizontally, they're often on soil that's damp and warm from the leaf cover. This exposes the plants unnecessarily to soil-borne diseases. Crops grown on a support have much fewer problems with rot, and therefore, waste. By allowing plants to grow up instead of out, you also limit their physical contact to neighboring plants. This is a major plus as plant diseases are readily transmitted through physical foliage contact.

And More … Is More

If these advantages aren't enough to get you scrambling for fencing and trellises, this just might push you over the edge: a bigger bounty. That's right, gardening vertically can actually increase your vegetable production. This increase in production is due to the plants and veggies receiving better air circulation and sunlight, which help maintain healthy foliage. Healthy plants with fewer pests and disease offer bigger yields, yet in a smaller space.

If you're growing a vining plant such as pole beans or squash up a vertical structure such as a trellis or another free-standing support, remember you usually have two sides of that support to work with. Voila! Double the production in the same amount of space.

Grown vertically, ripe veggies have a much better chance of being spotted by the gardener. One obvious advantage is that you don't accidentally pass one up only to spot it at a later date—after it's past its prime for the kitchen. But another good reason to harvest vegetables when they're ripe, but not overripe, is because for many plants an overripe fruit is a signal to halt production. Cucumbers, for example, will produce like mad until one or two of the fruits becomes overripe and left on the vine. At that point, as far as the plant is concerned, it has met its goal by producing fruit that contains mature seeds—ready for the next generation of cucumbers. Thus, production comes to a full stop. When the not-fully-mature-but-mature-enough-for-dinner fruit is popped off the vine, the plant keeps trying by maintaining production.

Something else less tangible plays into the higher yield and it has everything to do with your basic mind-set. Because vertically grown vegetables are easier to maintain (less time, less energy) you tend to not feel overwhelmed by a mid-summer, fully mature garden—and you actually tend it more. A little TLC goes a long way toward high production.

You'll also find that by growing vegetables up, you'll be able to fit more varieties into a smaller space.

Vertical Is Versatile

You can get creative with the vertical gardening technique by tapping into the unique versatility this style offers. Vining plants can take full advantage of the situation by reaching as high as they'd like for the sun, never running out of light (or room) like they can when grown horizontally.

As a bonus, vegetables that wilt in the blazing sun, such as lettuce, can be planted in at the base of a sun-worshipping climber like pole beans and relax in the shade. Veggies climbing up a vertical structure will also act as a windbreak for more delicate or young plants.

Other less obvious advantages of growing up instead of out are hiding or disguising a view. Camouflage a swimming pool pump or an open compost pile with your vertical garden. An air-conditioning unit or the place where the garbage cans are stored benefit from a vertical garden wall. Your plants can soften the view of the chain-link fence and also act as a privacy screen for your home. Remember, if those plants happen to be annuals (as many veggies are) then you'll only have a living screen for one season.

Some of you may have noticed that vegetable plants can be downright beautiful. If this hasn't crossed your mind before, consider:

- The brilliant red, yellow, orange, purple, white, green, and striped tomatoes
- White, purple, lavender, and striped eggplant
- The blues, grays, and greens of cabbages
- Red, green, purple, and all shades between of lettuce

If you've previously grown vining plants such as zucchini or squash in the traditional way, you'll find it interesting to note that instead of the leaves at the base of the plant becoming yellow, sparse, and scraggly, the bottom of the plant will be full of leaves. This fuller appearance will help convince you that veggies can be botanical eye candy, too! In fact, your vining and fruiting vegetables can actually be a surprising focal point in your garden landscape. Color isn't the only thing that adds to the view, but the shape, size, and texture of the plants, as well. Other upright crops such as espaliered fruit trees and grapevines offer many months of structural beauty in the garden, too.

Location Is Key

So, you know it's possible to grow food crops just about anywhere when they're grown vertically. You're probably also pretty pumped now that we've broken down all of the benefits that go along with "growing up." Before you start shopping and going hog-wild with the saw and screw gun, you'll want to know where your vertical garden and its structure(s) will be placed.

Logic says that you'll want the sunniest spot in the yard or around your home or apartment. But that's not a hard-and-fast rule. For instance, planting traditional sun-worshipping vegetables such as tomatoes, squash, corn, and melons in a sunny place is the right choice. But an area that receives only partial sun/semishade can fill the bill for certain crops such as the following:

Arugula
Beets
Bok choy
Celery
Collard greens
Cucumbers
Kale
Lettuce
Mustard greens
Peas
Potatoes
Radishes
Rhubarb
Spinach
Swiss chard

Some herbs will have no problem with light shade either, including the following:

- Basil
- Chives
- Lemon balm
- Mint
- Oregano
- Parsley

If you're very limited on space, then the places around your home might end up dictating which crops can be reasonably grown. The first thing you need to do is think about what you'd like to plant and then do a little research on their needs. Finally, poke around your home for the perfect spot.

UNDERSTANDING FULL SUN TO FULL SHADE

Light can change in all shady garden areas, just as it can with the sunny spots. For instance, you may have full sun or light shade under a deciduous tree in the early spring, but you may have full shade under that same tree in the summertime. There are many charts that describe all of the various levels between full sun and full shade, but here's a general guideline:

Full sun. Although it's not a hard-and-fast rule, gardeners refer to full sun as 6 to 8 hours of direct sunlight a day—or no less than 7 hours.

Full shade. Approximately 3 hours or less of sun a day is referred to as full shade. There may be some direct or dappled sun there for a while, but when there's less than 3 hours, it's considered full shade. The best plants for this area are those that will thrive in the shade as opposed to those that can take part shade. Most veggies and herbs won't do well here.

Part shade. Part shade receives 4 to 6 hours of sun a day, either direct or dappled. You'll often find great morning sun here; however, by lunchtime (or soon after) the area is shaded. You'll also find this under open-branched trees or trees that are trimmed high, which allows good light.

Light shade. Light shade varies; it often gets brilliant morning sun with a dappling of shade in the afternoon. If you live in a zone that receives intense sun, this may also be treated like it's a full-sun area.

As you can see, it's not an exact science; it never is when nature is involved. I've noticed that 2 hours in either direction doesn't make much difference unless we're talking about those vegetables that need a good sun-soaking, such as tomatoes, melons, and pumpkins. To fully mature and produce well, plants like these truly need a full 6 to 8 hours in the sun.

Once you've found the perfect planting spot, you're almost ready to dig in (so to speak), but there are a few more pieces of information that will lead you down the successful path on your vertical venture. By the way, the following guidelines apply to all types of gardening, both vertical and horizontal.

Start with a "Caveman Sun Blueprint"

So where is the full sun in your yard? I'm guessing that you said, "on the southern side of my house." Theoretically, you're correct. But, in reality, you might also have structures such as another building, tall trees, or fencing that's shading an otherwise sunny spot. No matter what kind of gardening you do, it's a good idea to get personalized information about the space that you'll be planting.

This includes one of the most important aspects: where you can find full sun. You can see how extra information can be useful to a vegetable gardener in particular. Let me be clear: you won't always want full sun. Some plants such as lettuce won't tolerate full sun for very long.

Finding the full sun areas in your yard or garden answers more than just one question. From there you'll understand what's happening in the areas surrounding the hottest spots.

One of the easiest techniques for finding full sun that I've come across over the years is one I've christened the "Caveman Sun Blueprint." It creates a very basic guide, but it remains my favorite to this day. All that is necessary to make a Caveman Sun Blueprint of your yard is paper, a pencil, and some time—specific time.

While I wouldn't hesitate to create this blueprint of my yard at any time, the best time would be in the spring on a day of blue skies (no clouds). The first thing you want to draw on your paper is a basic sketch of your home and property from a bird's-eye view. You'll want to add all the significant things such as your home, large trees, walls—even a neighbor's fence just for site reference.

You'll draw circles on your page at 8 A.M., 12 P.M., and 4 P.M. At 8 in the morning draw the first circle, which will include all of the space in your yard that has sun at that hour. At noon, draw the second circle to include all of the sun at that time. At 4 in the afternoon, you'll make the last circle in just the same way.

See the area on your blueprint where the three circles intersect? This is the place in your yard (or garden) that receives the most sun during the day. It's prime real estate for tomatoes, pumpkins, melons, peppers, beans, etc.

This rough-yet-handy drawing is certainly accurate as far as where your particular yard (or area of the yard) has the most sun. However, it doesn't tell the full story. For instance, it doesn't tell you how long that sunny spot stays sunny, right? So that place on your blueprint where all three circles intersect may only have that sun for a few hours each day.

In this case, even your brightest place may not be enough for those veggies and fruits that need 6 to 8 hours of a good sun-drenching. The simplest way I have found to figure it out is just to glance out the window every hour and log how many hours the sun shines on that space. Do this on the same day that you're creating your Caveman Sun Blueprint—you're committed to a few hours anyway.

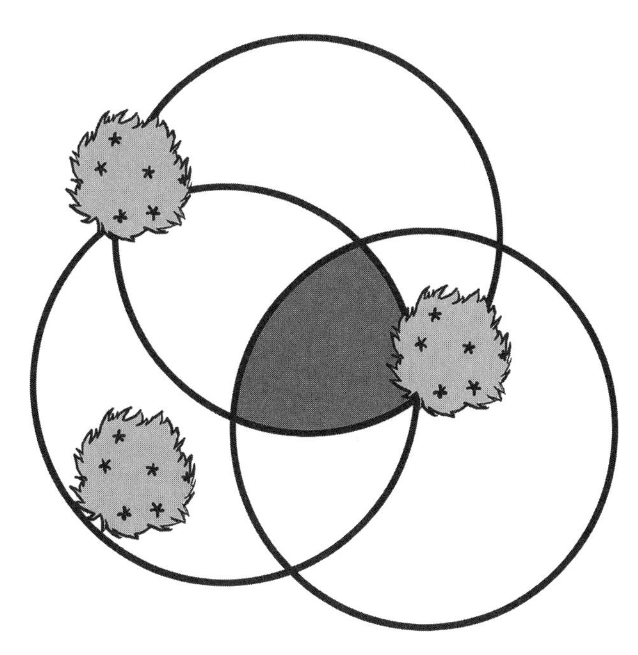

The simple Caveman Sun Blueprint will show you areas that get the most sun.

Know Your USDA Growing Zone

Growing zones are sometimes touted as a very big deal in the gardening world. I'm going to tell you up front that I'm not overly zone-focused. Still, one must start somewhere and I have to agree that it's a good place to begin. What kind of gardener would I be if I didn't explain zones and why they're good as a basic, general guide? Just remember that like everything worthwhile in life, there's so much more. Let's get to the why, what, and how of growing zones—and then I'll tell you why *your* place is completely unique and unlike anywhere else on this map.

The United States Department of Agriculture (USDA) has growing zones outlined and numbered on a Zone Hardiness Map. The numbers on the map reflect the lowest temperatures in a given area. Plants are assigned numbers according to the lowest temperature they'll tolerate and then you cross-reference them with the map numbers. In this way, you'll have a fairly good idea which plants can live in your yard and not freeze to death.

Hardiness numbers for plant varieties can be found in plant and gardening books, on seed packets, or on the plant's container. If the container isn't marked, there should be a plant tag tucked into it somewhere.

You may have heard that the zone map is only helpful for planting perennial plants, trees, and shrubs. It makes sense because the map is showing the lowest temperatures, which are obviously in the winter months. Some suggest that the map is moot when discussing vegetables since most of them are treated as annuals. All you really need is the right amount of warm days for that vegetable variety to fruit and be harvested.

Still, I think that it makes sense to have some basic ideas about where your garden stands, and the zone map is the first place to start. Your zone number comes in handy for describing your general area to other nonlocal gardeners and, of course, most vegetable gardeners will end up growing fruits (which are usually perennials) at some point, as well as ornamental plants.

It's not hard to find your zone number. In fact, online, there are several sites where you can type in your zip code and the zone number is generated in nanoseconds: planthardiness.ars.usda.gov/PHZMWeb/Default.aspx. If you happen to have a couple of gardening questions, you can get all the local information that you need by contacting the Master Gardeners at your local Cooperative Extension office.

If you're looking at a printed map, there are some versions where you'll notice that each zone is divided even further into A and B regions. This breakdown reflects a 10° difference between zones that are otherwise very close in temperatures. Also, be aware of the date on the map; the USDA put a brand-spanking new one out in 2012.

DOWNER

There's a map out there that has the potential to confuse people called the Sunset Climate Zone Map. Originally, Sunset's map was created to break down climate zones even further for the gardeners in the Western United States. It's still worth looking at. But you have to realize that these maps rarely line up. For instance, on the Sunset map your area may be a solid zone 7, while on the USDA map, you'll be a 9a. Don't let any of this throw you, though. It's not a deal-breaker if you're still confused.

The Magic of Microclimates

This is where I explain that your garden is special and no one else's climate is exactly like yours—even if it's in your neighborhood. As I hinted at earlier, the USDA growing zones are general guidelines that are basically accurate. However, no map can guess at the specific microclimate that's around your city, neighborhood, or yard.

So, what exactly are microclimates? They're specific, local atmospheric areas where the climate differs from the larger, surrounding area (such as the numbers of USDA zones). Your general zone is affected by individual circumstances, be it natural (such as wind) or man-made (such as the southern side of your home). These are your microclimates—and you can play with them.

Think of microclimates as zones-within-zones. Personally, I think this is where gardening gets fun and takes on a life of its own (so to speak). Because, technically, you may not be able to control the weather; but you can manipulate and adjust—often significantly—your growing zone by using (and making) your unique microclimate.

For instance, a friend may have told you that a certain plant variety doesn't survive the heavy winter frosts in the area. It may be confusing when that same plant is surviving winters just fine in your yard. If your zone isn't acting as your zone "should" chances are that you've planted it in the right microclimate. Maybe it's in a bed against a wall that has a southern exposure. That side may stay just warm enough to avoid a killing freeze.

Naturally Occurring Microclimates

Topography plays a major roll in creating microclimates. If your garden is on top of a hill, it's going to get more wind than one situated lower. Wind is very drying to both the soil and the plants, so choose varieties accordingly, or adjust your watering schedule. Drought-tolerant plant species would do well here, or those that can tolerate a windy beating.

Are you gardening in a valley? Frost is more of an issue in valleys than at higher elevations because cold air is heavier than warm air and settles in the basin. Plants that like moisture will do best here.

Which side of the hill you're on makes a difference, as well. A northern slope receives the sunshine last and is slower to warm up. While the south-facing slope warms up faster, it can be a mixed blessing. If a late frost hits spring-blooming plants, they could suffer a setback. Other microclimate factors are rainfall, soil type, and large bodies of water. Lakes, rivers, and oceans will moderate the air temperatures of any nearby inland areas.

Manipulating Your Microclimates

Whoever said, "You can't fool Mother Nature" never met a gardener. We can and we do as often as we can get away with it. While anyone, anywhere, can use their microclimates, urban and suburban gardeners usually have a few more man-made structures to work with and can take the greatest advantage of that situation.

Gardeners can take advantage of the very things that would otherwise seem to be in the way: walls, houses, and neighboring buildings. Permanent structures can have a huge effect on the immediate area surrounding them. Buildings can act as wind barriers or conversely, create wind tunnels.

Walls made of brick, stone, cement, or stucco will absorb heat and radiate it during the cool night hours. Walls not only hold heat effectively, but they can also provide shelter and be a protective windbreak for plants.

GOOD TO KNOW

If you're using the side of your house or a wall to grow sun-worshipping vegetables, remember to use any wall but the north-facing one, as they won't get enough sun there to produce well. A southern exposure sees the longest hours of sun; a west wall will get the intense afternoon sun; the east wall will have morning sun, and a (true) northern wall will receive no direct sun.

Using the sun's exposure can also be the difference between a perennial plant wintering-over or not—even if it isn't supposed to survive the cold months in your zone. Bougainvilleas, for example, don't usually make it through Northern California winters like they do in Southern California. But I've seen bougainvilleas that are alive and well in Northern California because they were planted against a wall with a southern exposure.

So if you're planting heat-loving veggies, be sure to plant them on the south side of your house—or any other type of wall. A word of warning about using a southern exposure as a microclimate: the sunniest side can also be the most drastic side for perennial plants because of fluctuating temperatures (think: a cycle of freezing and thawing).

Conversely, if you're looking for a cooler place, perhaps for lettuce, then go for the north side. The north side of your house might also be the best place for early flowering fruit trees like cherries and peaches. A late spring frost will set fruit production back, and the idea here is that if fruit trees are planted where there's a northern exposure, it can help suspend blossoming until the frost date has passed.

By the way, because morning sun is gentler than the sunshine throughout the rest of the day, plant tender plants on the eastern side of any structure. Save the brilliant west side for your sun-worshipping melons and pumpkins.

Although we tend to think of a southern exposure as a hot spot, it isn't always the sizzling area that it could be when there's a structure such as a neighboring building or large tree situated between your planting space and the sun—and there we have yet another microclimate. This is fun, right?

Consider too that walls have a downwind and an upwind side. Use this to your advantage by remembering the upwind side is the right place for water-loving plants as it's going to receive more rain than the downwind side. Plants growing on the downwind side will be protected against a driving rain. This can be an excellent little microclimate to have around.

Other examples of creating microclimates are mulching practices, paved surfaces, fences, balconies, and rooftops.

Garden Structures as Microclimates

If you've used all the natural microelements around your home, as well as the permanent man-made structures such as buildings and walls, you can always practice a little climate control by creating it with your own hands. There's no end to the garden elements you can create.

Cold frames and hoop houses. Think of cold frames and hoop houses as very short green houses that, as a rule, don't have any heat source. They're most widely used for getting a head start on vegetables in the spring or to extend the growing season beyond summer and into the fall or even winter. In fact, in climates with mild winters, some cool-weather crops (such as kale or lettuce) can be grown all winter long in a hoop house or cold frame.

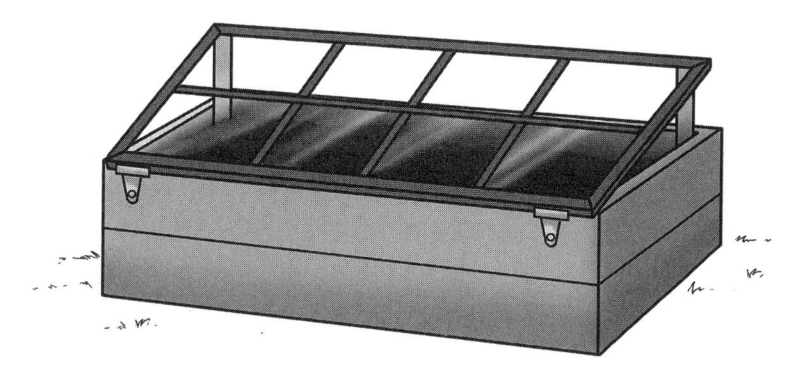

A cold frame extends the growing season so you can plant earlier and harvest longer.

The difference between cold frames and hoop houses has more to do with the construction than the function. Cold frames are built as framed, box-type structures that have no bottom. They usually have a lid or door on the top, which can be raised to different levels. To take full advantage of the sun, the lid is made of glass or clear plastic. This allows for excellent heat collection while keeping frost off the plants at the same time.

You can also use cold frames to toughen up (harden off) vegetable seedlings that started their lives under lights indoors. It's the go-between place for baby plants from the protective house to the exposed garden bed.

This versatile garden tool can be constructed as a portable structure or permanent, depending on your needs. I think that the quaintest cold frames are those made from recycled old windows. Or set one up using a discarded glass door (and frame). If you haven't had the pleasure of utilizing a cold frame before, I promise that once you do, you won't want to do without one ever again.

GOOD TO KNOW

Make a cold frame in a hurry using six straw bales and window frames. First make a rectangle with your bales by placing two bales side-by-side on one side, then at the end of those two, place another bale to make the end cap into an *L* shape. Repeat this for the other side and the last end cap. Place your plants into the middle of the bales and lay the window frames over the top—instant protection!

Hoop houses are often built with a flexible material to hold a cover such as PVC pipe, which is bent over an entire garden bed, and then attached. Several of these hoops are placed at even intervals all the way down the bed. In order to protect the plants in the bed, say from a surprise frost, you would simply toss some plastic sheeting or plant fabric over the hoops. In general, hoop houses aren't the halfway house that are cold frames. Plant roots remain undisturbed and you can take immediate advantage of the good soil that you've created in the garden beds by getting the plants in sooner.

Hoop houses make it easy to toss a cover over the garden during a cold snap.

Another way to create hoop houses is by making a frame-and-cover structure that lifts off the garden entirely and is then replaced as needed. Last, here's where you can see the looseness of the terminology. Greenhouses are often constructed of hoops and plastic (complete with door) and may be referred to as hoop houses or greenhouses—either moniker is fine. I tend to look at it this way: if I can walk upright in a hoop house, it's earned the title "greenhouse" in my book.

Row covers. Although row covers are even shorter than what we think of as hoop houses, they perform the same way. Row covers are used within a garden bed over specific rows of vegetables as opposed to covering the entire bed. They're useful for keeping out insects while still allowing water through, and the fabric helps warm the soil, as well.

Sometimes a heavy gauged wire is used to hold a lightweight synthetic plant material over the row of plants. Other times a "floating" row cover is used in which the fabric is placed lightly on top of the plants with the bottom edge of the fabric secured with small rocks or boards.

Cloches. Cloches are used to protect individual plants and seedlings from the spring frost. Many things can be utilized as a cloche; back in the 1800s, cloches or bell jars were made of glass. These old-time cloches were pretty, functional, and practical—especially in terms of small, home gardens. If you seek them out, you're sure to find some of these gems for a fair price, but you can make your own, too. You can also use things that you happen to have hanging around the tool shed. A terra cotta flower pot turned upside down and placed over a sturdy seedling makes a simple cloche. Add some straw inside to act as insulation. Or recycle your plastic water bottles and milk containers by cutting off the bottom and placing them over tender seedlings. Remove the cap from the top of the container for ventilation.

DOWNER

Don't forget to remove the cloche as the day grows warmer! Cloches can protect seedlings, but they will wilt them just as fast if it becomes sauna-like inside.

Cold frames, hoop houses, row covers, and cloches all help create wonderful microclimates. Raised beds, pots, and containers are other great examples of climate manipulation. The soil in a raised bed heats up faster than the soil in the earth, which gives seedlings a head start in the early spring, as well as lengthening the growing season in the fall. So don't let the number on a zone map have the last word on your garden plans. Get clever and manipulate your growing zone by using any microclimate resources you can find and put them to work for you.

The Greenhouse

Now we've come to the ultimate in climate control: the greenhouse. If you can get your hands on a greenhouse (any greenhouse) you'll be all the better for it. Heating inside the greenhouse is preferred to keep plants alive through the winter. Don't worry about heat, though; greenhouses are used successfully even when they're not heated—I've used a "cold" greenhouse (without heat) for many years. But when there is a heat source, a greenhouse offers the gardener incredible versatility.

Even those without artificially heating can have some temperature regulation, from making use of automatic vent openers to let hot air out and fresh air in, to attaching bubble wrap to the inside for winter insulation.

The first image that may pop into your head when you hear the word "greenhouse" is the hothouse type filled with fancy tropicals. It is, indeed, one great way to take advantage of the little building. But greenhouses are useful for so much more:

- Start your spring and summer vegetable and flowers seeds in the greenhouse. It's the perfect nursery for your baby plants until they're ready for a permanent spot in your garden.
- Just like a cold frame, you can harden off seedlings that you've started in the house in the greenhouse. Don't forget that this is a halfway house; the young plants will still need an adjustment period between the greenhouse and the outdoors.
- Grow some cold-loving vegetables in the greenhouse during the winter. Many vegetables such as lettuce, carrots, radishes, and peas will carry on through the winter for fresh food year-round. Whether you'll need to add a little heat will depend upon how cold your winters get.
- If you have a heated greenhouse, you can also winter-over your houseplants there. Why would I have my houseplants in the greenhouse? Because a lot of my houseplants spend the warm months vacationing under the cover of my back porch. Come fall, I find that I don't have nearly enough places (where there's a good light source) to keep them all in the house.
- If you enjoy propagating (making more of) your own perennial plants, greenhouses are excellent for housing cuttings that are waiting to take root.
- If you like to keep "parent" plants around for winter or spring propagation, this is the right place to hold them over.

Greenhouses range from a basic utilitarian unit to stunningly ornate; from slightly larger than a breadbox to larger than my entire house. Depending on their budget, most gardeners choose the more modest versions.

My greenhouse is only 6' × 8' with fiberglass glazing (the see-through material on the sides), complete with automatic vent opener that lets in fresh air and allows hot air to escape. This sturdy little greenhouse runs $600 to $800. The larger the greenhouse, the better the glazing (such as glass); and more bells and whistles will put you in the $1,500 range and up.

The simplest, least expensive greenhouses are basically shelves with a plastic cover that slides over the entire unit. They're approximately 62" tall, 27" wide, and 19" deep. But this depends on the manufacturer. You won't be able to walk upright in these little greenhouses, but there's usually a zipper so that you can reach inside and arrange plants and seedlings. They're priced anywhere from $65 to $120.

The next size up is 4' × 4' or 5' × 5' with a big plastic cover that slides over the top. These have shelves on two sides of a walkway and you can unzip the "door" and step inside. They're priced from $150 to $240 and up.

YOUR MOST VALUABLE TOOL: THE GARDEN JOURNAL

Any day is a good day to start a garden journal. Still, deep winter is perfect because you can start at the very beginning of the growing season and warm up to keeping record of your gardening experiences.

Useful journal materials include the following:

Binder. I like the kind with the clear window slip on the front for decorating purposes.

Lined paper. Record planting dates and plant descriptions; describe them in detail, including how tall they'll be, what color, and so on.

Graph paper. Use it to draw the shape of the garden or areas in your yard. Just eye it—don't fuss over correct scale. It's art time; there's only room for the creative brain in the garden journal.

Blank paper. Keep your drawings here, or cut and paste ideas from magazines. Tape pictures of your garden and plants, both the successes and failures. If you have favorite quotes or sayings, sprinkle those throughout your garden journal.

Colored pencils. You'll need the biggest assortment of colored pencils that you can find, because you'll be drawing. Oh, yes, you will. The idea is to give your memory as many reminders as possible about what went on during the growing season. Draw what's in your garden anywhere and everywhere in your journal.

Paper pockets. Use the pocket dividers or whatever strikes your fancy; they come in handy for receipts, pictures, and seed packets with seeds still inside. You can also tape empty seed packets inside the journal so you can refer to them during the growing season.

Garden Journal Records

The single most helpful thing in a garden journal are the dates that you write down. It might be the date you planted the perennial or the date you planted tomato seeds or harvested fruit. But the dates are going to be your biggest helpful hint for planning next year's garden. Especially if you're a vegetable gardener, dates will also tell you if you've chosen the right vegetable variety for your growing zone, as well as the frost dates for your specific zone. Also jot down the following information:

- Pests you've found hanging around.
- Pollinators that are around.
- Notes on other wildlife you may spot in your yard or garden.
- Description of the soil. Good loamy stuff? Or is it a work in progress? Raised beds or pots?
- Record of how your plants did. Were you happy with the varieties that you chose? What would you like to try next time?

After you've added notes, pictures, drawings, clippings, and extra information to the pages, you'll notice that your garden journal isn't just about your garden. Far from it. It becomes part of the timeline of your life. It keeps the memories of what was important to you that particular garden year.

Best Bets: Consultants

If you're not completely satisfied with the information you've gathered through your own research, there are gardening resources and experts all around you who are more than willing to lend you a hand.

My favorite resource for consulting about plant varieties, growing zone questions, and pest or disease troubleshooting is the Master Gardeners office. Master Gardeners are graduates of a volunteer program (Extension Master Gardener Program) who are trained to advise and educate the public on gardening and home agriculture. Here in the United States, they're affiliated with the land-grant university and the Cooperative Extension offices. They receive a thorough gardening education and because they have access to university materials, are extremely efficient researchers. They're truly your best bet, and can be reached directly through the Cooperative Extension office in your county.

My next favorite resources are local nurseries and garden centers. What's the difference between nurseries and garden centers? I admit that the terms are often used interchangeably, but there's technically a difference. Nurseries actually grow plants that might be directly sold to the public, landscapers, and garden centers. Sometimes nurseries sell to all three, and they may have some other garden products for sale, too. Garden centers order the plants that they sell to the public from nurseries. They not only carry plants, but they offer a lot of other garden-related products, as well. You'll often find containers, fertilizers, seeds, books, and garden ornaments at garden centers.

This isn't to say that some garden centers don't grow some of their own plants, and nurseries may also sell fertilizer and the like. It can be a fine line, but a nursery's primary focus is to grow plants. I've also found that I can get some in-depth advice and helpful tips the from nursery staff who actually grow the plants.

Vertical Garden Structures and Containers

2

Gone are the days when only those with a giant square of land, complete with neatly plowed rows and engine-powered machinery, grew their own food crops. As people moved closer together—deeper into suburban and urban areas—those who were determined found a way to have fresh, homegrown (and chemical-free) food for their kitchen tables. They actually found many ways—and they did so with gusto.

Those who were handy with wood and saws found endless ways to create support structures for their vining crops. Recycle hobbyists collected discarded items and grew vertically challenged plants in the most unlikely containers. Stories spread about people who were growing enough crops in their own backyards to supplement their family's meals all summer—and through some of the winter if they happened to also be canning and freezing their bounty.

It turns out that one of the most beloved and successful crop-producing techniques is vertical gardening. Traditional structures such as boundary fences, trellises, lattice, arbors, and obelisks were gardeners' natural first choice for growing vining vegetable crops. After that, things only got better. Companies saw such an interest in small-space gardening that they developed containers and structures to help gardeners' endeavors to grow food anywhere at any time. Garden products available now range from planters, hangers, and containers to instant garden beds, stands, and pockets, and transportable trellises.

The first thing I talk about in this chapter is *how* vegetables (and other plants) climb. I also point out some of the usable things that may be already in place in your yard. Then I get into the non-climbing vegetables and some products on the market that allow you to grow those vertically, too.

Twiners, Scramblers, and Clingers

The first image that usually comes to mind when we think of "vertical" vegetables and fruits are the types that climb up a fence, trellis, or other secured structure. Indeed, this is certainly the beauty of the vining crops such as cucumbers, the-other-squash (butternut, acorn, pattypan), gourds, pumpkins, melons, tomatoes, zucchini, peas, kiwifruit, berries, and grapes. And all kinds of beans will climb, including green (snap beans), shelling (drying beans), soy bean (edamame), fava (broad bean), and lima (butterbean). Last, but never least, are the climbing flowers such as sweet peas and clematis.

All of these plants can be grown up onto vertical structures because they "climb," right? Or do they? The truth is that some actually *do* climb as a habit, yet others are just sort of posing. Still, no matter how they get up there, taking advantage of the vining vegetables and fruit clearly pays off in spades.

The Twiners

Let's start with the true climbing plant group. It includes pole beans, peas, pumpkins, cucumbers, zucchini (and other squash), melons, kiwifruit, grapes, sweet peas, morning glories, and honeysuckle vines. Authentic climbers can achieve this in one of two ways: by twining or tendril. Twining plants will wrap their stems around anything that stands still long enough. They twine either clockwise or counterclockwise depending on the species.

Some climbers have curly little side shoots off of their stems called tendrils. Tendrils reach out and wrap themselves around the nearest support, too. It's simply two methods of achieving the same end. Climbing veggies and fruit are the right produce for growing against structures that are outfitted with a material such as a chain-link fence, railing, rope, netting, or wire mesh.

Tendrils will wrap themselves around the nearest support to pull themselves upward. (Photo courtesy of Anthony Deffina)

The Scramblers

The scramblers or leaners are open, prostrate shrubs pretending to be climbing plants. Raspberries, blackberries, climbing roses, and bougainvillea all belong to this group. Leaners often (not always) have thorns that don't actually help the plant climb, but by securing them to whatever they're leaning against, prevent them from sliding back.

Scramblers shoot up walls, trellises, and other supports, often cascading down the other side of the structure. If you want to maintain control of scrambling plants, you need to offer some mechanical support. Regularly tying the growing stems to a support and a bit of intentional pruning is all that's necessary.

The Clinging Climbers

Clinging climber plants such as Boston ivy and Virginia creeper make their way up walls, concrete, stone, or brick by literally sticking to structures with teeny suction cups. Other clingers like climbing hydrangea and English ivy have grabbing, aerial rootlets on their stems for support. This group is made up of primarily *ornamental* plants as opposed to vegetable plants. But for the sake of rounding out the list of climbing types, I thought I'd mention them.

Ornamental and Permanent Structures

What should your plants be climbing? Well, the most logical place to start is with the foundation or permanent (or semipermanent) structures in your yard or garden. Fencing that you already have on your property, trellises, or arbors that were previously used for roses, clematis, or other ornamental plants are all fair game as a support for vertical vegetables and fruit.

Boundary Fences

The fencing that separates your property from your neighbor's is a logical place to start when planning on gardening up. One way to utilize solid, wood-paneled fencing is by attaching hanging baskets, pots, and other growing containers to it. With this type of vertical garden, this wouldn't be about vegetables that climb, it would be about veggies and fruit that grow well in containers such as lettuce, radishes, strawberries, herbs, and the like.

A collection of garden pots can be mounted on a wall or fence, which not only takes advantage of the vertical space on the fence, but also frees up the ground beneath for, well … more pots. There are many ways to get flower pots up onto a fence or perhaps a deck rail. One way is to slide your flower pots into basket-shaped wire mounts. The wire baskets have a hook that fits right over a fence or railing. Special wall clips can be screwed into a wall or fence and then the pot is simply attached to the clips. Also on the market are circular wall sconces that are attached to the wall; the flower pot slips right into the metal ring. Other great products are wire grid (like a cattle panel), which comes with pre-attached circular holders; the entire grid is then mounted onto the wall.

Some companies offer special containers that are specially made for wall plantings. These pots or baskets are flat on one side so that they lie flush against the surface. They come in ceramic, plastic, terra cotta, and iron.

Deck rail planters are typically mounted in one of three ways: attached directly to a flat, wood rail; as a hanging planter with an arm over the rail; or as a sleeve in which the bottom of the planter "fits" over the flat rail. Variety abounds here, too, with plastic, wood, and iron basket types.

Wood fencing made with flat panels doesn't offer anything for climbing veggies to "grab" in order to hoist themselves up, but attaching a trellis or other grid-type material is a game-changer. Now, the tendrils and twiners have all the support they need. (See Chapter 4 for more about support structures.)

Some yards are surrounded by low, picket-type fencing. Fences that are constructed with both vertical and horizontal materials can be used as a home for climbing veggies. Hanging planters can also be secured to the top of this type of fence and still be within the gardener's reach.

Chain-link fencing, while technically ugly, turns out to be undeniably useful as a vertical garden structure. It's a climbing plant's dream. Not to mention that when something with full foliage is planted against it, it's transformed into an attractive "solid" wall. I've grown grapevines along our short, chain-link fence and I loved the look.

DOWNER

Most vegetable plants are annuals, which means that the plant is going to die at the end of every season. Which translates into you'll be picking the dead, brittle, twining vines off and out of that chain link. Capische? Now, this information never stopped *me,* but I felt it was only fair to offer up the dirty details. By the way, the grapevines weren't nearly as bad as the dead green bean vines.

Solid cement wall fences have an additional function that most don't: a flat, secure top. Rectangular planter boxes can be placed along the top of the wall for an instant vertical garden. Plant those boxes or troughs with peas and harvest the pods as they grow *down* instead of *up.*

Vining vegetables make good use of unattractive chain link and quickly cover the fencing.
(Photo courtesy of Annie Haven)

Trellises and Lattices

Trellises are a fast and easy solution for an upright garden structure. The terms *trellis* and *lattice* are most often used interchangeably——and that's pretty accurate. But let me explain the subtle difference. Lattice or latticework actually describes a type of pattern that's on some trellises. It's usually made of thin wood or metal that creates a diamond or crisscross decorative pattern.

It becomes a trellis when the lattice is used (framed or unframed) as a support for plants. Peruse any garden center, nursery, lumberyard, or hardware store and you'll find trellises with rectangle, square, and fan shapes, as well as the traditional diamond pattern. Of course, some trellises don't have a pattern at all.

Lattice is often painted white or a blue-wash color. Many wood trellises are left unsealed and the gardener can seal or paint it themselves. The wood can also be left raw, especially if it's made with redwood, cedar, or teak, as these woods deteriorate slowly. If you're looking for a carefree material, look for those made with wrought iron or heavy plastic.

As far as climbing materials, please don't stop at prefabricated lattice. Almost anything climbable can be used as a trellis. Other clever structures that can be used as trellis are ladders, old gates, fences, and doors. Willow and dogwood branches pruned off of trees can work, too.

Those of you born with a creativity bone may decide to design your own trellis. And if you're not sure where to start, you'll find some easy-to-follow directions for you in Chapter 4. Plus, in Chapter 5 I get your imagination juices flowing with ideas on recycling and repurposing.

Arbors and Arches

An arbor (or arch) is a plant support that goes over something such as a path or gateway. They can be large or small, thin or wide. For the purpose of this book, we won't be discussing these large, substantial structures, but rather, their petite cousin the arbor or arch.

You'll recognize arbors as having an arched, half-dome, or a flat top. Designs can range from cottage, to contemporary, elegant, or Victorian. You'll find that many of them have lattice within the framework. They stand anywhere from 6 ½'—to 9' tall at the center of the top. Gate entrances and the beginning of a garden path practically beg for one. Arbors can be used as entrances to various garden "rooms," or to create a cozy, out-of-the-way spot for lovers to nibble on dangling fruit.

GOOD TO KNOW

What about painting arbors and arches? Many people paint them white or green, which certainly adds a finished look to the yard, especially if you choose the same color as the trim on your house. Remember that painted wood requires a little maintenance. After several seasons, it'll chip or peel and you'll need to scrape the paint, sand it down, prime, and then repaint it.

A portable arbor can be placed in the center of a garden bed, or with one end in one raised bed and the other in a neighboring bed, emphasizing and embellishing the walkway in the center. If you'd like to use one at the garden entrance, you can place two planter boxes at either end, which will give you a new vertical garden space in an instant.

Plant cucumbers, grapes, small melons, beans, peas, gourds, mini pumpkins, indeterminate tomatoes, and nearly any vining vegetable or fruit at the base of your arbor or arch. Don't forget that planting some annual climbing flowers such as nasturtium, Black-Eyed-Susan vine, hyacinth bean, morning glory, or cup-and-saucer vine alongside you're veggies can add some extra beauty to this structure.

Why flowering annuals? Remember that with a few exceptions, most vegetables are annual plants (or grown as annual plants) and will die back at the end of the season, then are replanted year after year. Even in the case of grapes and cane berries, these plants are pruned, so you wouldn't want to disturb perennial or shrubby plant types such as climbing roses.

Obelisks

These handy and often beautiful structures are free-standing towers with a pyramid shape. Obelisks are placed either over the top of a climbing plant, or young plants are planted next to each leg for easy reach. Their construction style ranges from plain, simple structures to works of garden art, and are used for ornamental climbers as often as they are for vegetables.

Simpler designs are made of wood with thin wooden slats on the sides. Some are made of wrought iron with copper caps on top while others are made of metal with decorative scrolling running up the sides.

An obelisk is the reason that I first thought it was perfectly fine to have a vegetable garden in the front yard. I'll plant a garden on the front lawn any day of the week, but an obelisk almost makes it mandatory. Your vining plants could find nothing classier to climb.

Pots and Planters

We know that veggies and fruit climb in various ways, making them a shoo-in for the group name "natural" vertical veggies. But what about the other tasty foods that we want to grow vertically? Don't they get a cool horticultural group name like the others? They certainly do; we call them the "vertically challenged" group and this includes carrots, lettuce, onions, chives, garlic, chard, kale, cabbage, potatoes, strawberries, and herbs. And don't forget the bush varieties of the otherwise climbing plants such as bush beans (as opposed to the pole types).

Vertical gardening isn't just about planting veggies that climb, it's also about choosing plants that will grow in containers—and anything else we can get our hands on. But first things first. For gardeners who don't have the time or, perhaps, aren't into construction of any kind, the market is your oyster. Following are some of the latest and greatest innovations from companies that make gardening a cake walk for those of us who want to grow *up*.

As you peruse the internet, your local gardening center, and magazines, you'll find even more creative products, as this market is growing daily. I thought I'd give you a place to start by sharing some of my favorites with you. (Check out Appendix B for more information on these products.)

Pots are the first containers I turn to when I'm considering a patio or deck garden. There's no shortage of choices, either; you'll find terra cotta, plastic, wood, glazed stoneware, metal, concrete, stone, and fiberglass. You could throw a rock in any direction and find an average planting pot on the market. But there are those designs that deserve a second glance.

Stackable Pots

Pots and other containers that stack are the perfect way to get the non-climbers up into the air. The Akro-Mils Stack-A-Pot Stackable Planters are multi-tiered planters made from UV protected plastic so they hold up year-round outdoors. This stackable pot system goes together in layers and disassembles easily for storing during the off season.

This stackable planter makes for a simple vertical herb or vegetable garden.
(Photo courtesy of Akro-Mils)

Stack-A-Pot is available in a mini size (14 qt. for $25), and a larger size (30 qt. at $35). It's perfect for crops such as herbs, lettuce, and strawberries.

Woolly Pocket Wallys

One of the latest (and greatest) recent planting crazes is the pocket containers. This particular wall pocket is made by Woolly Pockets and is made to last, using 100 percent recycled plastic bottles. Wallys have a moisture barrier (in some of the styles this is military grade) and the outside has a breathable felt that allows moisture to evaporate. At the inside back of the pocket is a "tongue" that's made to wick water down to plant roots and keeps the soil evenly moist.

Woolly Pockets can make a full garden right on your fence!
(Photo courtesy of Shawna Coronado)

They're no harder to hang than a picture, and come with the hardware to attach them to masonry, drywall, or sheetrock. You can get a surprising amount of soil in those pockets! Wallys come in black, brown, blue, tan, and green and range in size. The smallest is 8" × 13" and costs $18, while a single original is 15" × 24" and costs $40. After that, the Wallys are constructed as connected pouches. The three-pocket Wally is $100 and the five-pocket is $150. There's also a place in the longer versions for drip irrigation lines. Nonclimbing vegetables and herbs are perfect for Wallys.

EarthBoxes

At first glance, this popular product looks like any common, rectangular planter box. But it's actually a pretty impressive little growing system in its own right. One of the interesting features is that there's a reservoir at the bottom of the system that keeps the soil evenly moist as long as there's water inside. A long tube runs from the bottom up past the top of the box for watering directly into the reservoir. Before you plant the box, there's a cover that's placed over the soil bed, which helps retain moisture, as well. You then cut holes in the cover and add your plant starts.

I found that there was very little concern about keeping the plants in the box watered. I filled the reservoir about once a week even in the dog days of summer. As the plants begin to fill out and produce fruit, you may find that you'll need to give it a drink a little more often; but not nearly as much as traditional pots or containers. My vegetable plants just went crazy in here—the EarthBox is a performer for sure.

The EarthBox has accessories that you can add, such as wheels (which I love), but my favorite add-ons are the stake-and-net system and white trellis; either one will turn the EarthBox into a vertical gardening planter that's perfect for climbing vegetables. The basic unit is priced at $33.

Mobilegro Portable Garden Cart

Here's a new product on the market that will not only pack your vertical space with crops, it'll wow all of your friends and neighbors, too. This extra-large, vertical planter has powder-coated steel and offers both functionality and beauty—on wheels!

Another nice feature is the built-in watering system that allows you to customize the settings for each pan. Mobilegro is available in custom panel designs, colors, and trim. The containers are 11.5" deep.

Currently available on the market (more styles are in production) is the three-tiered cart, which is 39" tall and has a 9-square-foot capacity, for $399; and the *t*-tiered cart, which is 50" tall and has a 12-square-foot capacity, for $499.

Mobilegro is like the Rolls Royce of the vertical container world.
(Photo courtesy of Angela DiMaggio)

Hanging Planters and Baskets

Hanging planters are one of the best ways to make good use of vertical real estate. There are endless ways to go about it, too. The eaves under the roof of your home are good places for hanging planters and baskets. You can also purchase metal "arms" that attach to the outside wall of a house or backyard fence as hangers.

GOOD TO KNOW

Hanging baskets often end up under the eaves of the house or the porch where there can be a lot of shade. If your baskets end up on the shady side of the porch, choose vegetables that do well with less sun, such as lettuce, spinach, arugula, radishes, oregano, parsley, cilantro, chives, garlic chives, lemon balm, mint, and bush peas and bush beans.

If you don't like the idea of attaching anything to the house, you can use iron rods called shepherd's hooks that are curved on one end. These are free-standing, as the bottom of the hook is pushed into the ground.

Any vegetable that can be grown in your average pot or container can be grown in a hanging planter or basket. Lettuce, peas, beans, strawberries, eggplant, herbs, small peppers, and cherry tomatoes all take well to suspended planters.

Lettuce doubles as an ornamental when paired with flowers in a hanging basket.

I know, I know, I talk about lettuce a *lot.* But since I've discovered the incredible variety of greens that are available, I've become obsessed—plus, lettuce and many other leafy greens are some of the easiest food crops to grow. One thing is for sure, shallow-rooted lettuce takes exceptionally well to being grown in hanging baskets and other sky-high planters. It's entirely possible to have a diverse lettuce garden suspended in mid-air.

Hanging baskets come in some wonderfully unique and attractive styles. My favorites are those that are made with a wire skeleton frame and then lined with a coco material. The material acts as a barrier between the wire and the plants, keeps the soil inside, and helps retain moisture (depending on the type of liner). If you look around, you'll find more than just the traditional bowl-shape, such as conical and oblong. Pay attention to which kind of liner you'd like if you choose a wire basket–type hanger.

Some liners will be thick and are meant to be planted at the top of the basket only. I enjoy the thinner material that allows me to poke my scissors through and push little plant starts anywhere in the basket.

I enjoy planting this way because the result is a bog "ball" of hanging plants; it looks wonderful done in all lettuces. Another variation on the same theme are hanging planters made of plastic that have planting holes built into them. Along with traditional pot-shaped styles, there are rectangular ones, too.

Get more crops out of hanging planters by attaching two or more together.

Another effective way to take advantage of hanging baskets or planters is to attach two or three in "stacked" fashion, one under the other. One excellent example is the hanging gutter garden.

Non-climbing vegetables can be grown vertically in a hanging gutter garden.
(Photo courtesy of Jayme Jenkins)

Upside-Down Planters

We now come to the infamous upside-down planters. What started as a terrific idea quickly became discouraging. The initial product of this type utilized a bag that was too small for most standard tomato varieties, although one might have success with a cherry tomato variety. What do I mean by too small? Well, large tomatoes like to get their roots into soil that's a little deeper than what is in the bag, and during hot summer months, it's difficult to keep a decent amount of moisture in there. For some reason it also troubles me to watch the plant struggle and twist itself upward to reach for the sun.

Let me be clear: I'm not saying that you *can't* grow something in this bag, but when it comes to tomatoes, I just don't know why you would *choose* this particular product. You can do so much better.

In the spirit of not throwing the baby out with the bathwater (after all, it's a pretty cool concept), these products have improved. There are upside-down container products on the market today made of quality materials and with more soil capacity. In any case, one of the best solutions that I've seen for planting upside down is to create your own containers with 5-gallon buckets, which is the largest soil capacity I've seen yet. I'll tell you how in Chapter 5.

Plant Stands

A plant stand can mean different things, and thankfully, every one of them is helpful as far as vertical gardening is concerned. The first type of stand is one built with graduating shelves or steps that holds whatever containers you decide to put on it. There's a bazillion types in any number of heights and sizes. In fact, before you run out and purchase new ones, this is one thing that shows up at garage sales, Craigslist, and Freecycle regularly. Never get rid of a plant stand—you're going to need it sooner or later.

Another kind of plant stand is the type that has the planters built right into it. It's a free-standing unit that may have two, three, or four rectangular containers attached to it. Something like this can become an entire vertical garden unto itself. Picture a three-tiered stand with strawberries planted at the top, lettuce in the middle, and peas planted in the last container.

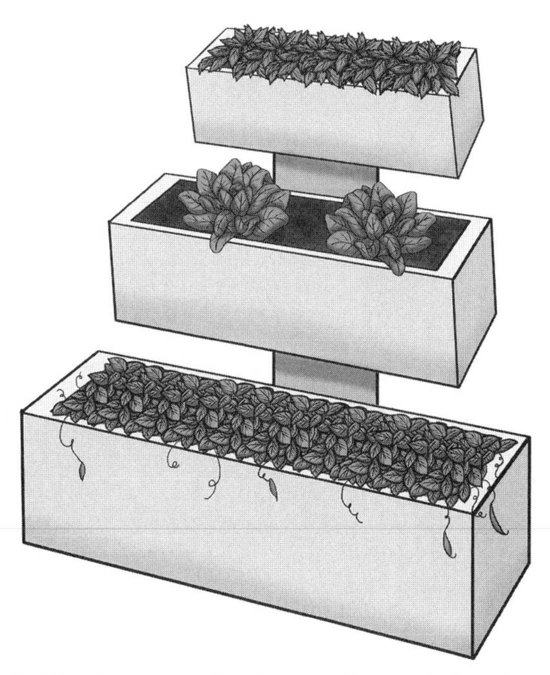

A tri-level plant stand makes a perfect vertical garden.

Pre-Fab Climbing Support

For vining vegetable plants in containers, you'll want to be on the lookout for structures that they can either twine themselves around or that you can attach their stems to. Fortunately, you don't have to look much farther than your local nursery, garden center, or the internet. You'll find pepper and tomato supports that fold up for easy storage, portable trellises, miniature obelisks, and a pole-and-net setup that resembles a small volleyball net.

Make Your Own Bed 3

Vertical gardens are created using the same gardening principles that should be practiced with any type of garden, but there can be some differences. For instance, as I talked about in Chapter 2, you might be using miniature garden "beds" such as containers like pots, rail planters, hanging baskets, or stackable containers for nonvining vegetables and fruit.

On the other hand, you may be gardening in a raised bed or a traditional in-the-ground garden bed where you'll place any upright structures for the climbing plants.

Of course, container gardens can be moved around if you aren't happy with their first positions on the porch, deck, or yard. Be sure to take some time to consider which crops you'd like to grow and how much sunlight is necessary to grow them before you build one of the more permanent raised or in-ground beds.

The Benefits of Raised Garden Beds

Raised garden beds are probably the greatest invention since facial tissue. You can grow vertical veggies and fruit in any bed available, whether it's a traditional plot, prefabricated instant beds, or something that's been recycled. You'll notice that when gardeners plant a vertical garden from scratch, it's often in a raised bed of some kind. Even a container or a pot is a modified raised bed.

GOOD TO KNOW

I'm often asked how deep I make my raised garden beds. The truth is, probably not as deep as you think. I've planted in beds that were 6" all the way to 24" and I can honestly say that I've never needed any bed over 12" deep. My preference is 12" beds for plants such as tomatoes that like to sink their roots deep, but 6"- to 8" beds for most other vegetables.

Raised beds that are loaded with fresh, loamy soil offer a bit of growing magic.

Extended growing seasons. Out of all the great benefits, this one is at the top of my list. Raised beds warm up earlier than in-ground beds, which give you a jump-start on the growing season. The soil also holds heat longer, so vegetable plants remain productive longer into the fall.

Soil control. One of the greatest things about raised beds is that *you* control the soil type and blend. Prepared garden soil and compost have a basically neutral pH, unless you've purchased a soil that's amended for acid-loving plants such as blueberries. This also makes it easy to alter the soil for certain plants in one bed without messing with the other. Plus, it starts out weed-free!

Excellent drainage and water retention. Actually, the magic here is that when you build a raised bed, you're often adding prepared garden soil (whether it's bagged or what-have-you). Prepared soil blended with compost typically has good tilth (porous and crumbly; resembles coffee grounds) and that means high organic matter. This soil holds water well, yet is free-draining as far as excess. So raised beds rarely become waterlogged, which gives roots room to breathe.

Reduced soil compaction. In a traditional garden bed, it's hard to keep your feet off the soil. This is important because plant roots need air and busy feet compact the soil. Raised beds should be designed to let the gardeners reach to the middle of the bed from two sides, thus your feet remain outside the bed and on the path.

Easier garden maintenance. Raised beds are physically closer to your hands making weeding, harvesting, and any other maintenance a simple task. There will be less weeds than you're probably used to—much less weeds. You'll save time, money, and labor on all your gardening tasks (amending, fertilizing, watering, etc) because you're not broadcasting any materials—they can be applied specifically to the raised bed.

Less or no soil erosion. Unlike the unframed ground, the soil and amendments you've added to a raised bed for the most part stay put through watering and winter rains.

More vegetables. That's right. Especially when we're growing food crops, we want as much as we can get. Why do we get more vegetables by growing them in raised beds? Well, it's not a matter of one part in this case, it's due to a little of everything above. And because you don't have to step between traditional rows, you can plant your veggies a little closer together. I'm not suggesting overcrowding your vegetables, but you can certainly plant a bit more intensively in a raised bed. Also, the leaves will shade most of the bare soil, which tends to smother out weeds—a huge plus in my book.

Gardeners report that they see one-and-a-half to two times higher yield in their raised beds than their in-ground beds. Now imagine the harvest you'll see from setting up vertical structures in your beds *and* gardening up!

If you've never planted vegetables in raised beds before, I promise that once you do, you'll be spoiled for the rest of your gardening life. That's a risk you might be willing to take.

Considering Raised Bed Materials

Even if you've never built anything in your life, I encourage you to have your first project be a raised garden bed. They're one of the easiest and most useful structures around your home. Look around your home because many of us have most of the materials needed to construct a bed hanging out in the garage or storage shed.

When it comes to potential framing materials for raised beds, there are a lot of choices. Like most things, it comes down to preference. But there are a few things that should be considered before making your choice. Lumber is the number one choice for many gardeners, so let's talk about wood first. Back in the day, chromated copper arsenate or CCA–treated lumber was quite popular, until someone figured out that the arsenic, chromium, and copper the wood contained was leaching into soils.

Since then, the Environmental Protection Agency (EPA) has stopped arsenic from being used in the lumber. Pressure-treated wood is now deemed to have no ill effects on people or animals and is processed in a more ecologically friendly way. Alkaline copper quaternary (ACQ), copper azole (CA), and micronized copper quaternary (MCQ) are some of the current wood treatments; arsenic was dropped from the formula.

DOWNER

Although wood is no longer treated with CCA, the new versions are now higher in copper. Both ACQ and CA's high copper levels have the potential to be a hazardous waste when they're disposed of. If you choose to work with chemically treated wood, your best bet is to wear a mask while you're cutting to avoid the dust particles. And if you have any waste, don't throw it onto a burn pile; take it to a landfill instead. If you're leaning toward using a treated wood, MCQ is said to be the most environmentally friendly and the least likely to leach into soils.

If you'd like to play it safe and skip treated wood altogether, there are some great alternatives:

- **Raw lumber:** You'll pay a little more for raw wood than the treated kind and it won't last as long. You're trading that for peace of mind. If you go this route, look for cedar or redwood because they have natural rot-resistant properties.
- **Cut logs:** Talk about cheap materials! I've built garden beds using logs for the perimeter. Surprisingly, they hold up for quite a few years. In fact, after 3 years the only part that showed signs of breaking down was the outside bark—not bad for free raised bed materials.
- **Composite lumber:** Depending on the company, composite lumber may be made of polypropylene and wood fiber or other recycled materials. Some can be extremely long-lived if they have UV-ray protection, as well. These materials are typically more expensive than raw lumber, but they last longer.
- **Recycled plastic:** Free and recycled materials are always a score. "Boards" made of recycled plastic is a win for the planet and a win for you because it's another long-lasting product.
- **Cinderblocks:** Okay, I realize they're not as attractive as wood. But, they're inexpensive, last forever, and go together in a hurry. We like this. There's a plan for making a cinderblock raised bed later in the chapter.
- **Concrete or concrete blocks:** If you enjoy working with concrete, it can be poured and shaped into a garden bed. If you have no clue how to work with concrete, there are several different styles of concrete and interlocking blocks that go together in no time. Both can be permanent.
- **Rock:** Although building a raised garden from rock may take the most time to construct, it's also one of the most beautiful permanent beds. They remain my favorite.
- **Brick:** Beds made of brick offer a different look than rock, but they add character all the same, and last forever.

The Safety Speech

Please don't skip this section. It may seem boring, but this stuff is important. Most of this section is common sense, but sometimes we get used to going about what needs to be done and tend to skip some (or all) of the basic safety precautions. We all need a reminder every now and again, so here it is.

Proper gear and attire:

Heavy gloves. Building can cause play all kinds of havoc with your hands. Cuts and splinters from wood, sharp tool edges, and prickly plants are all everyday injuries that can be prevented with a pair of work gloves.

Rubber gloves. In order to avoid chemical burns and other hazards, disposable, heavy rubber gloves are a must if you're using chemicals or other toxins.

Goggles. Whatever you do, please, please use safety goggles when you're using any power tools, striking concrete or metal, and pruning trees and shrubs. I can't stress enough the importance of protecting your eyes at all costs—they simply can't be replaced.

Closed-toed shoes. I don't think people realize how important their toes are until they break one. If you have work or hiking boots, they're your best bet, but even tennis shoes or garden clogs are better than leaving your feet exposed with sandals or flip-flops.

Disposable face mask. Use them when you're sawing wood (sawdust) or using airborne chemicals.

Work precautions:

Power tools. If you've never used a potentially dangerous power tool such as a skill saw, don't turn one on until someone shows you how to use it properly. Keep all of your body parts away from a power tool's working parts (bits, blades, cutters). By the way, make sure any clothing or your hair is secured to ensure it won't pose a hazard. Electric saws have a tendency to grab whatever is put before them. If you've had several beers or Mojitos, please don't use power tools.

Read directions. This goes for everything, including manufacturer's instructions for using tools and chemical labels (organic or otherwise). The words are there for a reason.

Sharp stuff. My friends, sharp or pointy things like cutters, knives, saws, and the like don't belong in your back pocket. And screws shouldn't be carried around in your mouth. Instead, carry building essentials in an open tool tray with a handle, 5-gallon bucket, or tool belt or modified apron.

Look around you. Before you begin construction on anything, take a look at what and whom is around you. Get the big picture. Be sure that anything you saw off isn't going to fall into something else. People, especially small children, should be cleared of the work area, as should *all* pets.

Case in point: My husband was putting up iron fencing and leaned one of the panels against a tree ("just for a moment") while he went to retrieve a tool from the garage. My hens chose to walk by the panel, which just happened to fall over at that exact moment—right on top of my Rhode Island Red, Penelope. She died almost immediately.

THE TOOLS BY YOUR SIDE

In this chapter and the next, you'll find building instructions for beds and vertical structures. For the most part, you'll need the same group of tools for any one of them. Here is a list so you can go ahead and gather those tools now:

- Tape measure
- Level
- Shovel (square or spade)
- Skill saw
- Drill
- Assorted size drill bits
- Screwdriver bits (that fit into the drill)

Build a Raised Bed

This single-raised bed is a perfect foundation for your vertical garden structures. It's also one of the easiest garden beds to build; no formal carpentry skills are necessary. Like any garden, be aware of which vegetables you'll plant in this bed and how much light they will receive before you set it up.

Gather your materials:

Gloves

Safety goggles

Power drill

1 drill bit (for predrilling holes)

1 driver bit

12 3" wood (deck) screws

4 4' long, 2" × 6" boards made of cedar, redwood, fir, or pine (cedar and redwood will last the longest)

Cardboard or newspaper

Soil blend: ½ garden soil, ¼ compost, and ¼ aged manure yields approximately ½ cubic yard of soil

If you're building a double-raised bed, you'll also need:

8' long, 4" × 4" lumber cut into 12" lengths (to serve as corner posts)

4 more 2" × 6" boards

12 more wood (deck) screws

Assemble a single-raised garden bed:

1. At one end of each of the 2" × 6" boards, use the drill bit to predrill three evenly spaced holes.
2. Switch out the drill bit for the (screw)driver bit. Attach the four boards end-to-end by screwing the wood screws through the predrilled holes and into the cut end of another board. When all the boards are attached, you'll have a completed bed frame.

A single-raised bed can be assembled in less than an hour.

Assemble a double-raised garden bed:

1. To be sure that a higher bed is secure (especially after it's filled with soil), it's a good idea to add corner posts. So instead of screwing your screws into the end of another board, you'll screw them into the side of a 12", 4" × 4" corner post. Repeat this until all four boards are secured by posts.
2. Predrill one end of each of the remaining four 2" × 6" boards.
3. Take the four boards and attach them to the corner posts the same way that you did the first four.

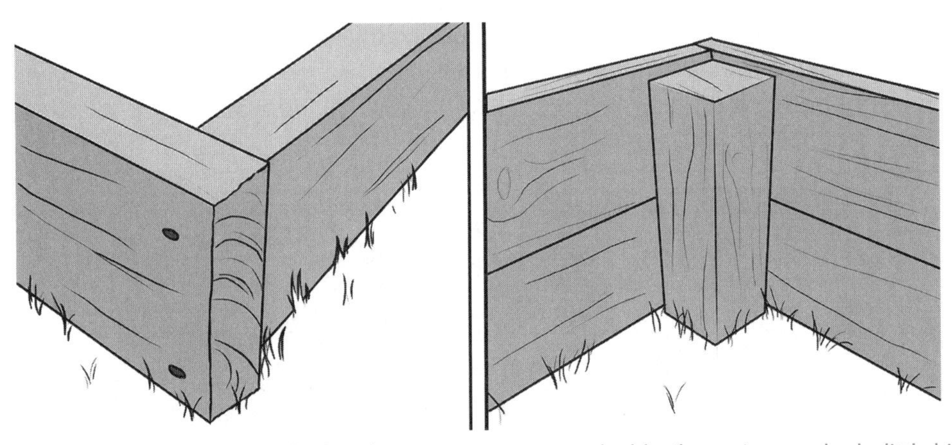

Left: The corner of a single-raised bed. Right: Use corner posts and add rails to raise your bed a little higher.

Once you've finished construction on either bed, line the entire bottom with cardboard or several layers of newspaper to smother any weeds that may decide to rear their ugly little heads. Your raised bed is now ready for soil, compost, and your vertical structures!

My favorite raised beds feature a ledge or cap that's been mounted to the top board of the bed. It sits like a triangle at the corners. I love these "sitting corners" for holding a giant glass of iced tea—and they're a big help during weeding or harvesting.

One last thought: be sure to build your raised bed only as wide as you can reach to the middle (from both sides). If the bed is up against a wall or fence, you should be able to reach to the back without stepping into it. You'll be thrilled with your forethought when it comes time to plant, weed, and harvest.

GOOD TO KNOW

Wood beds are typically joined at the corners with galvanized or stainless-steel screws. But keep your eyes peeled for some terrific products on the market that allow lumber to be joined together by using corner brackets. These handy devices are made so the boards slide the wood into the correct position. All without having to dig through your tool bucket.

Cinderblock Raised Bed

Cinderblocks are one of the most versatile materials ever and they come together as a garden bed in a hot minute—no drilling or screwing required (but don't forget to wear gloves!). This project is as easy as it gets, plus it's a simple task to make it bigger or taller to suit your needs.

The following directions are to complete a 4' × 5' cinderblock bed. If you prefer a deeper bed, stack a second row of blocks on top of the first, making sure to stagger the blocks to make the structure more secure.

Gather your materials:

Gloves

12 cinderblocks

Rake

Square shovel

Level

Small rocks, gravel, or sand

Soil blend: ½ garden soil, ¼ compost, and ¼ aged manure yields approximately ½ cubic yard of soil

Assembe a cinderblock raised bed:

1. Remove any weeds that are in the area where you're building your bed and rake it flat.
2. With your shovel, dig a slight depression into the soil all along the perimeter of the bed where you'll place the cinderblocks. Make these depressions for all four sides of the rectangle; there will be four blocks on the long sides and two on the short sides.

3. Place your blocks into the depressions with the holes facing up. Start with a long side and place four blocks end-to-end.
4. Make the next long side with four blocks.
5. The two short sides will have two blocks (holes up) which should be placed up against the inside of the last blocks at each end of your longs sides; thus, creating corners.
6. In order to have water drain uniformly out of the bed, you'll want to be certain that your bed is as level as it can be (without worrying about perfection) by placing the level on top of each side of the bed.
7. The holes in the blocks can be filled with small rocks, gravel, or sand for extra stability. You could also fill them with soil and use the holes as extra planting space.

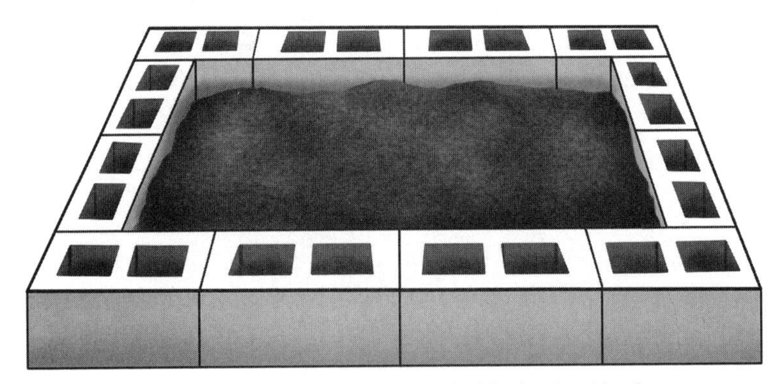

Anyone can put together a cinderblock raised bed.

If you're concerned about critters such as moles, voles, or gophers getting to your bounty before you do, this is the time to place chicken wire or another barrier such as landscape fabric at the bottom of the bed. Now you can fill your bed up to about an inch or so from the top with your soil blend. When I have taller beds, I typically don't fill the beds that close to the top, but this one is short and I like all the soil that I can get.

GOOD TO KNOW

Want to add supports to your bed to hold a row cover? All you need are four to six semicircular brackets, and four to six, 6" or 12" long (depending on how deep your bed is) of 1" diameter PVC pipe, and ½" screws. Secure the PVC pipe on the inside of the bed, 1' to 2' from the ends by screwing the brackets around the pipe and onto the bed frame.

If you've made a longer bed, add two more pipe pieces to opposite sides at the middle of the bed. You now have hoop holders on your raised garden bed. When you need a hoop house, use ½" to 1" diameter PVC pipes so that you can bend them into semicircles and slide into the holders. Then simply toss plastic over the top!

Instant Garden Beds

Instant garden beds offer some serious instant gratification. These market products pop or fold open, or come as kits that easily slide together within minutes of removing the packaging. As soon as you add garden soil, they're ready to plant—no fuss, no muss. I *adore* instant garden beds as they're handy for people who don't have the time or the inclination to build a bed from scratch. If you are completely new to growing things, you can't ask for a better first garden.

As far as price goes, there's a broad range with something that fits everyone's pocket book. Look for them online, or in hardware stores, garden centers, and big-box retailers (warehouse stores). I discuss some of my favorites in the following sections. See Appendix B for more information on these products.

The Little Acre

This is a great little pop-up garden bed "bag" made of industrial strength woven polypropylene. The Little Acre comes in two sizes, Original 3' × 4' × 12" and Junior 2' × 3' × 8". This instant garden bed comes complete with bottom liner, making a fantastic instant garden bed for patios, small yards, condos, apartments, or anyone who wants to create a little garden in minutes.

Because it does drain well, I'd set it on top of a pallet if it's being used on a wood deck or patio so the water can drain and dry freely without saturating the wood for long periods of time. Best of all, it's a simple task to place a trellis inside and grow your vegetables vertically.

When the growing season comes to an end, you can remove the spent plants and add some fresh compost or composted manure and it'll be ready for next season. Or remove the soil, hose off the garden bed, and store it. At $49 for the Original and $39 for the Junior version, The Little Acre is not only super handy but very affordable. This is the bed I'd give as a gift.

The Little Acre can give you an instant garden on your patio or deck.
(Photo courtesy of The Little Acre)

Greenland Gardener

Another terrific instant garden bed, in my opinion. The difference here is that the Greenland Gardener doesn't have a bottom, so this wouldn't be the one to use on your patio or cement driveway. However, one of the nice things about a bare bottom is that while you're adding and amending the soil that's filling the bed, the critters of the earth (such as worms) are being called in at the same time.

Any organic matter that you place in your Greenland Gardener bed will get worked by the macro- and microorganisms that are in your soil naturally—which means the soil directly beneath the bed will become better and better as time goes on.

Slide the Greenland Gardener raised bed together in minutes.
(Photo courtesy of Katie Elzer-Peters)

The kit comes with composite sides that slide into corner pieces, becoming a raised garden bed in literally minutes. Single beds run from $35 to $58, depending on the dimensions. Although it arrives unassembled, the only thing faster are the pop-up beds.

Woolly Pockets

Just like the Wally that I mentioned in Chapter 2, Woolly Pocket Meadows are made of 100 percent recycled plastic water bottles. But *this* pocket is an instant garden raised bed complete with a bottom liner, which is perfect for balconies, patios, and even cemented driveways.

Meadows have a moisture barrier all the way across the bottom and up the sides, as well as a water reservoir that allows plants to soak up as much water as they need. Breathable sides let excess moisture evaporate. The beds are offered in the following colors: green, brown, black, blue, and tan, and can be purchased in several sizes. Wallys are modular and are made to fit together, so you can always add on another later.

The smallest bed, Lil' Meadow, is 9" tall × 24" × 24" and costs $90. Meadow, the next size up, runs 9" × 48" × 48" and costs $200. The Deep Meadow is 18" × 24" × 24" and costs $250. They're all durable, attractive, and created to last. I would add a trellis and grow anything in these bad boys.

Woolly Pocket Meadows (raised beds) are great for growing just about anything.
(Photo courtesy of Woolly Pockets)

DIY Structures from Scratch

You're going to love this chapter; it's all about creation. All of the structures that you'll find in this section are easy for anyone to assemble. Use the directions as a jumping-off point for your own imagination.

I discuss various climbing supports for vining vegetables and fruit. I thought it was important to talk about climbing support options so that you can decide which one will work for you and your vertical garden. I also have thoughts about plant-tie materials for attaching climbing structures to their supports.

What I want you to take away from the first sections of this chapter is that the basic support system directions can be created the same general way, but you can swap out both the climbing and the framing material for whichever ones you'd like.

I also want to give a quick reminder to flip back to Chapter 3 and reread the section titled "The Safety Speech" and the sidebar entitled "The Tools by Your Side" before you get started on a project.

Summing Up Support

A support system is the combination of whichever material your plants will be climbing and whatever is holding that material in place (the frame). Before you commit to a support system (and purchase the materials) you'll need to consider the plants that you're growing. In the following sections are some thoughts on the materials that are used to create vertical support systems.

Climbing Materials

Lightweight climbing materials such as netting and twine are perfect for vegetables such as peas and green beans. But if you're planning on vertical melons, you'll want something sturdier, such as wire panels or wood.

Trellis netting. Nylon netting made specifically for trellising climbing vines is an excellent choice for everything but the heaviest fruit (like melons). It'll last for several seasons and although sizes may vary, there's always enough room to collect ripe fruit.

String, jute, or twine. String, twine—whatever you have hanging around the garage or junk drawer—break it out for your garden structures. You probably already know just how versatile these items are in everyday life, and it's no different in the garden. String it horizontally from bottom to top for light twining plants like peas and beans. Or for something with more to grab onto, weave strands horizontally as well to create netting.

Baling twine. Hay bales are often held together with colorful baling twine, as opposed to the old baling wire of the past. After the hay is gone, many people are left wondering how to repurpose the plastic twine left behind.

Secure one end of the twine at the bottom of a support system by wrapping and tying—and then stapling it—to the frame. String it up to the top and repeat. You could also make netting out of it the same way I explained with the jute or string.

Bailing twine needs to be secured at the ends—not just tied. If you tie it, the knot will hold well for a while, but will loosen fairly quickly. This is why I don't recommend it for plant tying in the next section.

Wire mesh panels (cattle or hog panels). These panels are made of galvanized welded steel rods and have square openings that vary in size. Those that are thick and galvanized are often referred to as hog or cattle panels and are my favorite metal mesh. Hog panels don't rust, and they last forever. Look for them at livestock feed stores.

There's also a thin, ungalvanized version called concrete reinforcing wire, which can be found at any home improvement store. These will rust (probably in the first year), but they're seriously cheap. A little spray paint will make you happier with them and they're easy to move and store.

GOOD TO KNOW

The thickest hog panels hold heat incredibly well. If this is the climbing material you're using, be sure to feel them during the hottest part of the day to be sure that they aren't getting so hot that they'll burn the new tendrils that are making their way up it. If you feel that the panels are too hot, add a shade cloth or umbrella in front of the trellis when the sun is blazing just until the plants begin to take over the panel. The leaves of the plant will cool things off after they fill in a bit.

Chicken wire. Chicken or poultry wire is tricky. On one hand, it's inexpensive and easy to handle—plus it's easily climbable for the lightweight veggies. On the other hand, it's *not* easily reusable because pulling the spent vines out of those holes is a nightmare. For many vegetables, it's a bad choice all the way around. It is simply too frail to use for something like a tomato vine cage; and for others, even if it did manage to stay upright, you can't get your hands through the wire to reach the fruit. It's really best to stick with supports that allow 4" to 7" of room for easy harvesting.

All of this said, I've used chicken wire in the past (and will again in the future) because I often have it around, it's an inexpensive material, and it works well for a temporary structure.

Plastic poultry fencing (a.k.a. garden netting). This netting is inexpensive and sturdy when attached to a frame, but the downside is just about the same as it is for the chicken wire, in that you can't get your hands through the holes. But as I've mentioned before, if I have something on hand that will work (even if just for the season) then I use it. In fact, this year I made a sandwich board A-frame (see instructions later in the chapter) and we happened to have the plastic garden netting on hand. So we attached it to the frame and are currently growing green beans up one side and cucumbers up the other. Although it'll be a little time-consuming, I'll more than likely pull the spent vines from the frame when the season is over.

Field fencing. Some people have what's referred to as "field fencing" left over from another yard project. It's a generic wire fencing that's used to border land and pastures. The spaces between the wire in field fencing are at least 4" wide and sometimes more, so you can easily fit your hands in the opening. You can find it at a home improvement or livestock feed store.

Chain link. If you're looking for a sturdy climbing structure it doesn't get much better than chain-link fencing and the metal poles that hold it up. It's one of my favorite vertical supports. The only drawback is that the holes may not be as small as chicken wire, but it can still be difficult to get the dead vines out of it once the season is over.

Lattice. Many gardeners find that lattice is a satisfying climbing structure that just happens to be attractive at the same time. The key here is to purchase latticework that's been well-constructed (don't go cheap if you want it to last). The key to using lattice for climbing plants is to match the thickness of the wood slats with the vegetable that you have in mind. A thin and inexpensive support is best with lightweight crops such as peas or green beans.

Bentwood (greenwood). Young or greenwood (branches) can be cut and used to create trellises, teepees, and arches by attaching thin branches in a row to a larger frame and then weaving more horizontally. There's no limit to the designs you can create. Willow branches are the most commonly used, but any bendable wood, such as elm, birch, hickory, and cedar, can take the place of willow.

Framing Materials

The second part of your support system is a "frame" to hold the climbing material in an upright (or slightly angled) position. Following are the most commonly used.

T-posts and rebar. T-posts are my favorite quickie supporting posts. Armed with them and a post slammer, I can't think of any structure that I can't build. Plus, T-posts are inexpensive … and yeah, also rather unattractive. Still, I can't get over their versatility in the garden and they remain my favorite.

PVC pipe. The cool thing about using PVC pipe is that you can make a ton of designs with it. It's sturdy, versatile, and available at good price, too. As far as aesthetics are concerned, some people like the smooth-white-plastic look, and others find it very unattractive.

DOWNER

From an environmental standpoint, PVC pipe (*Polyvinyl chloride*) isn't readily recyclable and its manufacturing by-products are toxic. That said, PVC pipe is labeled as safe to use as gardening structures. For more information, visit The Healthy Building Network at healthybuilding.net.

Galvanized pipe. Galvanized pipe offers structural support for a lifetime, but can also be removed and reused elsewhere. Because you can purchase elbows and other fittings, alternate sizes and designs are only limited by your imagination.

Lumber. Wood structures are the darlings of the garden, and it only makes sense. It's attractive, sturdy, and naturally blends into a garden environment. Lumber is fantastic for those who have permanent projects in mind. The most beautiful and long-lasting wood creations are those built from the heart of the wood. Redwood and cedar are popular for the long haul, and pine will have the shortest life span.

Bamboo. Framing made of bamboo is for lightweight climbing plants such as green beans, peas, cucumbers, and mini pumpkins. I love using bamboo because it's inexpensive and easy to handle. In loamy soil, two heavy bamboo rods can be pushed into the ground with a net strung between them. They're also my favorite material for teepee trellises.

Tie One On

Gardeners can always use ties for one thing or another. Sometimes it's to attach a climbing support—such as netting or fencing—to posts or poles. Other times we need ties to secure growing vines to their support systems.

In any case, ties come in handy, so the question is, "*Which* ties work best?" It's often just a matter of preference on the gardener's part. Still, there's some logic behind the right ties for the job, so let's take a look at your choices.

Twine, thick string, and jute. The uses for this type of tie are as numerous as the varieties and styles, so be creative.

- The good news: If you have it already, it's free; if not, it's cheap. And did I mention versatile? It can be used for guiding vines or securing a climbing structure to posts, etc. By the way, jute, twine, and cotton string are all organic fibers and therefore are fit for the compost pile once they show too much wear and tear after a couple of seasons.
- The bad news: If it's thin twine, you'll have to be careful not to tighten it too much when you're securing a plant.

Wire. There are a lot of variables with wire. Some is flexible and bends easily using only your hands. You'll need pliers in order to twist others. Don't overlook wire for securing *nonliving* materials (not plants). Wire is inflexible and will end up cutting into (and eventually killing) plant stems.

- The good news: It's strong, durable, and inexpensive. It's potentially easy to use, and if you choose a galvanized wire it won't rust.
- The bad news: If you need pliers to twist the wire, then it'll be harder to use (therefore, more physical effort). If it isn't galvanized, it'll rust and become unsightly.

Fabric strips. I'm talking about the fabric strips that you can rip up yourself from old sheets and the like. Fabric strips can be strong, so I would (and do) use them for tying anything.

- The good news: If you're repurposing old sheets or other fabrics, then these ties are 100 percent free. And as long as you don't pull too hard and crush the vine, they're gentle on plant tissue so they are great for tying plants. Fabric ties are pretty easy to reuse later.
- The bad news: Depending on the fabric, they may stand out in the garden (unless you're using a camo print ...).

Plastic plant ties (commercial). There are a few types of commercial plant ties that are staples in nurseries and garden centers. The first is the thin, green plastic type that comes on rolls. You'll find them hanging on an end cap by the tomato plants because they're handy for securing tomato vines as they grow. They're also the least expensive of the commercial ties.

Another variation of green ties is the Velcro ties ... which don't actually "tie" at all. They wrap around the plant and the structure and wrap back around to cling onto themselves. You may find other ties that work in much the same way with only a slightly different approach.

You may have already figured out that these products are for guiding plants to their climbing structures. It's possible to use them for other things, but they're not necessarily strong enough for a tight hold.

- The good news: Obviously, purchasing commercial products is convenient, considering you don't have to scour the garage for something that resembles a tie when you need it. They're also created to be visually appealing. In other words, you don't see them; they basically blend in with the garden. As far as being reusable, I suppose that technically they could be reused—especially the self-clinging types. They're durable products, especially the self-clingers.
- The bad news: In my opinion, when you cut the stretchable tape to size (because you're trying not to waste any), it's difficult to reuse that piece on another project. I'm not saying that it *can't* be done, just that you probably won't. For this reason I don't consider the stretchable green tape to be reusable. (Your mileage may vary.) The clinging type has much more reusable potential in my opinion. The stretchable green ties are cheaper than the Velcro-types, but compared to some of the other ties, they're still the priciest way to go.

GOOD TO KNOW

Vegetable plants that produce heavy fruits, such as melons, cantaloupes, spaghetti squash, pumpkins, and the like, will need to have a sling of some type to support the maturing fruits. Seriously large vegetables such as 25-lb. watermelons simply won't work grown vertically. On the other hand, if you're growing a very small vegetable such as Baby Boo pumpkins or 2-lb. melons, a sling won't be necessary. It's the veggies whose weight falls somewhere in between that will need a little help.

There are slings commercially available; however, you'll probably find several things around your home that would work just as well. Netting of any kind will work —bird netting, too. Also pantyhose, T-shirts, and any cotton fabric will do the trick.

Just tie one end of the sling to the trellis or whatever support you're using for that veggie. Gently slide the sling under and around the fruit and then tie the other end of the sling to the trellis. You may want to make the sling larger than the fruit is at that time. If it's made to fit the fruit at the size that it is now, then you may have to add another (bigger) sling later.

Metal fence clips. If you're working with T-posts, metal fence clips are worth their weight in gold. You'll need pliers to bend the ends back, but they're fast and secure.

- The good news: I love to use them for taking the place of wire as far as securing climbing supports to posts.
- The bad news: They have limited use.

Zip ties. We've all used zip ties at one point or another; they're considered the poster child for securing in a hurry. You wrap them around whatever you'd like and slide the narrower tail end into the hole at the other end and pull it through. Rigid plastic teeth have an unrelenting grip. Some police officers now carry handcuffs that work just like zip ties.

Zip ties should be strictly used for securing a climbing material to a post, or a post to a post, etc.—never use them to secure plants to anything. They'll end up damaging the plant in one way or another. If they don't scrape the tender vines, they bend and break them.

- The good news: I adore zip ties. I tie things with zip ties just because I can. They're fast, easy, strong, durable, and relatively inexpensive. You'll find them in various lengths and many colors. (I like green or clear in the garden.)
- The bad news: They're not reusable (clearly). It's not always simple to remove them from a structure if you need to. On the other hand, they're not always hard to remove, either.

Twist ties. Twist ties are for securing plants to their trellis or netting, etc. You'll find them packed up at the garden center or you can get them off of a loaf of bread.

- The good news: Well, the best news is that if they're coming from bread bags and such then you're repurposing; we like that. Repurposed, they don't cost you a dime and they're easy to use.
- The bad news: I usually have to tie a few bread bag ones together to make them long enough to wrap around anything. The plant twist ties sold new are made longer.

Plastic clips (commercial). Special hard plastic, plant clips are used to guide plants along their supports. If I had only one or two tomato plants, I could possibly see investing in these clips. The truth is that they feel like overkill to me. But it's great to have so many choices.

- The good news: Plastic clips work best with heavy-branched crops such as tomatoes. They're durable and certainly reusable.
- The bad news: They can be expensive (comparatively speaking).

Build a Sandwich Board A-Frame

Sandwich board A-frames are simple to build, store, and modify. If you're not worried about storing it, you can forgo the hinges and screw or wire the top of the structure together (where the boards meet). Using the information in the previous sections, you can switch out the plastic netting for any other climbing materials that you'd like.

This sandwich board A-frame was made with plastic poultry/garden netting because it's what I had on hand. Also notice the short, rectangle "planters" at the bottom. We had extra fence boards and added them on later to create a small bed for the shallow-rooted green beans.

Gather your materials:

6 4" × 6' fence boards

Jigsaw

6 5/8" wood screws

2 sets of metal hinges

16 1½" wood screws

1 3' roll poultry or garden netting

Manual heavy-duty stapler (T-50)

Drill gun with screw bit

GOOD TO KNOW

To prevent splitting while you're drilling screws through lumber, it's always a good idea to predrill (make pilot holes) in any piece of wood before you add the screws.

Assemble your sandwich board A-frame:

1. Take two of the 6' fence boards and using your jigsaw, saw them in half so that you now have four 3' boards.
2. Place two of the 6' boards vertically on the ground in front of you.
3. Take one of the 3' boards (that you cut) and lay it horizontally at the top of the 6' boards so that the ends of the 3' board lay over the top ends of the 6' board.
4. Secure one end of the horizontal top board to the top end of one of the 6' boards with two ⅝" screws. Do the same to the other board ends. This will give you one frame.
5. Using the remaining two 6' boards and the remaining two 3' boards, create another frame.
6. Make sure the 3' cross board is "on top" (meaning that the 6' long boards are pressed against the ground) and lay one frame down in front of you.
7. Take the second frame that you created and place it flat on the ground above the first one. Remember, the 3' cross boards should not be touching the ground.
8. Space the two hinges evenly apart on the 3' cross boards. Using two 1½" screws, attach one of the hinge flaps into one of the cross boards on the frame, and the other hinge flap to the cross board of the second frame. Repeat for second hinge.
9. Stand your sandwich board A-frame up. Unroll the garden netting partway.
10. Starting at the outside bottom of your A-frame, staple the end of the netting to the bottom of one panel.
11. Roll the netting up and over the other side of the entire frame all the way down to the bottom of the other side. If you're using netting that's much wider than your frame, simply use sturdy scissors or wire cutters to trim the material even with the sides of the frame.
12. Staple the netting to the bottom cross board. Add staples up all four sides of the A-frame about every 8"–12" intervals.

Remember to stand the frame up *before* you secure the netting (or whatever material you're using) to it. If you add the climbing material while it's flat on the ground, there won't be enough give in the material to allow it to bend into an A-frame. Your sandwich board A-frame is now ready for the garden!

MODIFIED A-FRAMES

A-frame structures are not only easy to modify to your personal style, but they're versatile in the garden. They can be placed inside a large garden bed, have rectangular beds attached, and they can be in two beds at the same time by placing the bottom of one side of the sandwich board into one bed and the other side into another. The center of the A-frame is over the walkway in between the beds.

Ladder A-frame. An A-frame that has horizontal boards in the center of the frames as opposed to mesh or netting is what I refer to as a "ladder" A-frame. This structure is excellent for supporting the heavier crops such as pumpkins, gourds, and melons. The boards offer strength for the heavier vines as well as a place to secure the slings that may be necessary for certain melon or pumpkin varieties.

A-frames that are used to uphold heavy fruit should be slightly shorter than ones used to support lighter crops such as Scarlet runner beans. Use screws to secure the long boards to the legs of the frame.

Wire panel A-frame. A couple of wire mesh panels can be set up in minutes if you use zip ties to attach one end of the panels together and then simply spread the other two ends apart to make a tent.

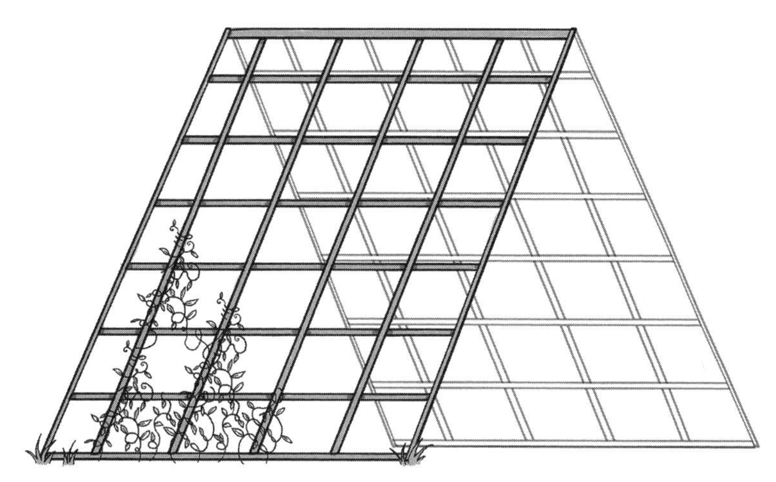

Probably the simplest A-frame you'll ever make and it can be closed up and stored for next season.

Double your vertical space by purchasing enough materials to create two more panels and set the A-frames up end to end.

Pallet A-frame. Make an A-frame out of two untreated wood pallets by standing them up on their sides and either wiring or screwing the top of the *A* together.

Usually, the pallets don't slip because they're pressed against the soil. But if you want to be certain to keep the frame at a certain width, you can attach eye screws to the middle of the sides of the pallets. Then measure wire that's the distance between them, adding 8" to 10". Tie the wire ends onto the eye screws.

A couple more thoughts on A-frame climbing structures: First, they can be made into cold frames by simply tossing plastic over the top and securing it to the wire or attaching 2' × 4' boards along the bottom of the plastic to hold it in place.

Another beautiful thing about this vertical style is that you don't have to waste the space underneath the frame. Use that area to plant crops that need less sun and less heat than the trellised crops, such as lettuce, kale, and other leafy greens.

GOOD TO KNOW

If you're making new raised beds, the beds should run north to south in order for both sides to get the same amount of sunlight. Place your vertical structures at the north side of the garden bed so the vining plants don't shade the rest of the garden.

Twists on Trellis

Only your imagination will limit you as far as the shape and size of a trellis and the materials you can use. Following are some tried-and-true trellis ideas ready for you to modify in your own way.

Make a Basic Trellis

This is a very inexpensive trellis that goes up in less than an hour. It's great for virtually all vertical vegetables, as long as you take the climbing material into consideration. Last year, I used a galvanized wire mesh panel and supported tomato plants with it. Here's how I did it.

Basic trellises are versatile and can be constructed as wide and as tall as the material you have on hand.

Gather your materials:

Post slammer or mallet

2 7' tall T-posts

Measuring tape

1 5' wire mesh panel (any width)

Zip ties

Assemble your basic trellis:

1. Using the post slammer or mallet, pound one of the T-posts 2' into the ground.
2. Measure the length of the panel from the first T-post to where the next T-post should go. Pound the second T-post into the ground until it's also sunk 2'.
3. Place the wire panel between the two posts and using the zip ties, secure one post to the side of the panel. Do the same for the other side.

A basic trellis can be modified to hold more plants. The corner trellis is built just like the basic trellis, but with another panel set into an *L* shape. I like to add this to a corner raised bed. It's also great to use as a "living screen" in front of a window that has little or no privacy.

Corner trellises come in handy for suburban yards and small-space gardens.

Post and Twine

This trellis will end up having the same look as the basic trellis above, but uses different materials. Using the string or twine as opposed to netting makes this project most wallet-friendly.

Post and twine trellises are a variation of the basic trellis theme.

Gather your materials:

Post slammer or mallet

2 8' long × 2½" diameter round wood poles

Measuring tape

Hand drill with screw bit

Screws or eye hooks (any size or diameter will do as long as twine can be wrapped several times around them)

1 roll of twine, heavy string, or wire

Assemble the post and twine trellis:

1. Use the post slammer or hammer to pound wood poles 2' deep into the soil. Measure 10' to 12' away from the first post and pound in the second post. You can make this trellis as long as you'd like to by adding posts about every 10' or so for stability.
2. Measure and mark the screw or eye hook placements so that they are evenly spaced on the posts. Add the screws to the wood at those marks. Do this for the second post (and the rest of them if you added more).
3. Attach your twine, string, or wire to the screws by wrapping the ends tightly and tying them off. All that's left is to plant!

Don't forget that any leaning type plants such as tomatoes will have to be secured onto this trellis every so often as the plant grows.

Bamboo and Twine

One could really get creative with bamboo and twine; the designs are limitless. Armed with inexpensive bamboo poles in varying thicknesses and lengths and a roll of twine or jute, the vertical growing world becomes your oyster.

Bamboo trellises are inexpensive, light, and easy to manipulate.

The bamboo and twine trellis pictured here could be modified by leaning it slightly backward and connecting yet another bamboo trellis to the top creating an A-frame. If you flip back to Chapter 1, you'll find a photo of another example of a bamboo trellis system.

Wire Panel Tent

Making tents with panels can give you a shelter in which to plant crops that need a little shade from intense sun. Once the vining plants make their way about halfway up the trellis, you can plant a leafy green such as lettuce underneath the structure.

Don't forget to take advantage of the space and dappled shade under this tent trellis!

Gather your materials:

2 7' tall T-posts or wood poles

Post slammer

1 wire panel (hog panel or concrete reinforcing wire; height and width of the panel or wire will vary)

Metal clips, zip ties, or wire

Tape measure

Pliers (if you're using metal clips)

Assemble your wire panel tent:

1. Measure the panel and mark this distance on the ground so you know where to sink the posts. You can also just place the panel on the ground so that you can see where to sink the posts.
2. Using a post slammer, pound the posts about 2' into the ground.
3. Lift the panel upright (short side up) and lean it onto the top of the T-posts. Secure the posts to the end of the panel with the metal clips, zip ties, or wire.
4. Place the other end at an angle into the garden bed (raised or not).

Flat Ladder

Flat wood trellises are practical to have around, plus they're easy to build and transport throughout the yard. Before you purchase any wood, look around for scraps from previous projects or ask your neighbors what they might have lying around their house.

Flat ladder trellises are perfect for vegetables such as small melons and cucumbers.

Gather your materials:

2 6' tall, 1" × 2" lumber

7 4' tall, 2" × 2" lumber

1 box of 1¼" wood screws

Drill gun with screw bit

Assemble your flat ladder:

1. Lay the two 6' boards parallel to each other about 3' apart on a workable surface such as a garage floor or driveway. The ends of the boards should be lined up evenly.
2. Place one 4' board perpendicular to the longer boards with a 6" overhang at each side of the long (6') boards.
3. Using the drill gun, secure the 4' boards where they intersect each other on both sides using two 1¼" screws.
4. Place the next 4' board approximately 6" from the last one and continue the process with each remaining 4' board until you reach the end of the 6' boards.

By the way, there's no magic in the number of 4' boards or 6' boards. It's simply one version of a flat ladder—and yours is an original!

Teepee

Teepee trellises are best known as a support for crops such as peas or pole beans. It's a happy coincidence that they just happen to make superb kid forts as well. Teepees are stand-alone structures that are constructed of bamboo poles or other long, slender wood.

This simple, upright structure is perfect for all gardens, big or small. Plus they can be modified for placement in half wine barrels by using much shorter poles for legs.

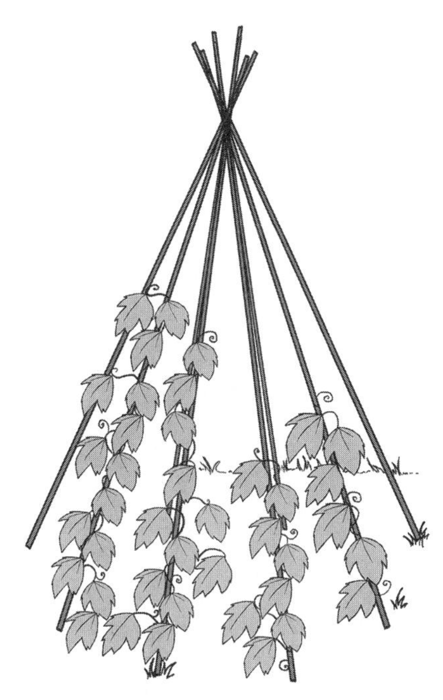

Have your kids help you construct a teepee trellis and then let them plant climbers such as runner beans and morning glories at the base of each pole.

Gather your materials:

6-8 6'–8' long poles (bamboo, thin scrap lumber, thick branches, etc.)

1 roll of twine, thin rope, jute, leather straps or wire

Assemble your teepee:

1. Place all of the poles evenly on the ground.
2. Using the twine (or whatever you have) lash the poles together about 1' away from one end. This will be the top of your teepee.
3. Stand all of the poles up at the same time and spread each leg out individually, keeping them evenly spaced. The structure should now be standing on its own, Indian teepee style. I usually do this step by myself, but it would be easier to have a helper!
4. You'll need to lightly secure the legs so it doesn't blow over. But if it isn't holding a heavy crop this is easily done be sinking the legs about 4"–6" into the soil.
5. If you're planting peas or beans, go ahead and directly seed 4–6 seeds around the base of each pole.

If you widen the space between two of the poles, you'll have an instant "doorway" for a kids' fort. If you'd like to grow crops that are a bit heavier such as gourds, pumpkins, and cucumbers, you'll want to add more climbing material. One great way to do this is to add twine, rope, or heavy string to the sides of the teepee.

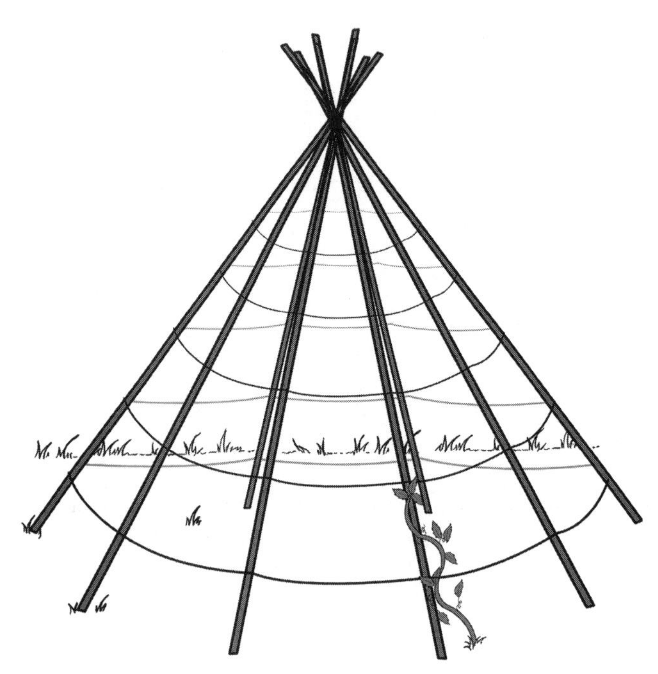

You start by tying the twine onto one of the legs and then bringing it onto the next pole and securing by tying or using the staple gun. Continue in this fashion all the way around the teepee until you reach the pole you started with. Start at the same pole, but move up 12" and do it all over again. Make as many horizontal lines as you'd like.

Wire Panel Arbor

All you have to know to make this sturdy trellis is the location of your nearest livestock supply store. Although it goes up in a snap, if you didn't need two people before, you will for this one. This arbor not only offers plenty of climbing space, it's capable of providing support for anything in your vertical vegetable garden.

Wire arbors make extremely strong trellises!

Gather your materials:

Post slammer

1 16' × 3' cattle or hog panel

4 5' T-posts

16 metal fence clips or heavy-duty zip ties

Pliers

Assemble your wire panel arbor:

1. Using the slammer, pound one T-post into the ground about 1'.
2. Measure 3' from the first post and pound the next post into the ground.
3. Lay the panel onto the ground and have your partner stand on one of the short ends.
4. Go to the opposite short end and lift it up as if you're going to bring it to meet the short end that your helper is standing on.

5. Step on the panel so that you create a permanent bend in the center of the panel. You can also create a couple more bends on each side of the center "crease" to make the top rounder.
6. Stand the panel up so that the creased area is at the top; this will create a tunnel or arch.
7. Place one short end against the T-posts that are in the ground. Using pliers, bend the metal clip ends securing the panel to the T-posts.
8. On the opposite short end of the arch, pound the remaining T-posts into the ground approximately 3' apart (or 4', depending on how tall you'd like the arch), the same way you did with the first two. Secure with metal clips.

Lifetime Tomato Cages

Here's a great way to construct some super-duper-sturdy tomato cages that'll last forever. We usually use concrete reinforcing wire because they turn out sturdy; we like brute strength around here! And the openings are hand-sized for easy fruit harvesting. Just realize that you can always use any sturdy fencing material that you have on hand.

You'll have to find the perfect place to store them during the off-season, but after you've used lifetime tomato cages, you'll turn your nose up at the flimsy store-bought types.

Our lifetime tomato cages are always built by two people. This wire likes to roll back up on itself while you're trying to work with it; when wire snaps back and hits you, the whole day is ruined. A helper makes it safe, simple, and much less frustrating.

Gather your materials:

A 5'–6' length of concrete reinforcing wire (5' will make a 1½' cage; 6' will make a 2' cage)

Small bolt cutters

Pliers

Assemble your lifetime tomato cage:

1. Have your helper stand on the short end of the wire while you roll it out.
2. If you purchased a roll, then use the bolt cutters to cut a 5' or 6' piece off of the roll. Leave some of the strands long (past the 5'–6' piece) to be used to help secure the ends later.
3. Put on work gloves and bend the wire into a cylinder shape; your helper can assist.
4. Now that it's formed into a cylinder, use the wires that you left sticking out to bend over the wire at the opposite end in order to attach them and make a continuous circle.
5. Make sure that the wire you've bent is pointed toward the center of the cage. Also cut off any other protruding wires to avoid getting cut while placing the cage or harvesting tomatoes.
6. Because you've been working with the ends, the cage may not look nicely smooth and round. Just shape and form it with your hands until it looks presentable.
7. The last thing you want to do is look at the very bottom of the cage. Do you see the very last circular wire that's touching the ground? Cut that horizontal wire off of the cage so that you have the vertical wires as stakes on the bottom.

Slip the cage over a tomato plant and press the wires into the ground for a little stability. I don't usually need anything else to secure the cage (it's a really strong and sturdy wire). But if you'd like to add something, just take short wood stakes and drive them into the ground up against the wire. Use thinner wire or twine to tie the cage to the stakes.

Cinderblock Planting Wall

Planting walls are terrific for vertically challenged and trailing plants. In fact, some climbers can double as trailers such as vining peas. Create this vertical wall planter in minutes for your porch or patio.

Cinderblock planting walls are best for plants that aren't particularly heat sensitive, such as herbs. These walls can hold the heat!

All you'll need is about seven cinderblocks and soil. You just "plant" the holes that are in the cinderblocks. Feel free to construct the wall into any shape that you'd like; just make sure that the cinderblock below the one that you'd like to plant is turned so that the holes are facing out as opposed to up. This block becomes the bottom of the hole for the cinderblock that's placed on top and is planted. Use the drawing above for guidance.

Terra Cotta Tower

This is a vertical system that's also for nonclimbing plants and adds some fun to the yard. It's great for herbs, strawberries, lettuce, and dwarf vegetable varieties. My wonderful friend Theresa Loe shared with me her instructions for making a terra cotta tower.

Gather your materials:

Post slammer (used gently) or mallet

1 6' rebar rod

6 8" terra cotta garden pots

Assemble your terra cotta tower:

1. Using the post slammer or mallet, pound the rebar about $1\frac{1}{2}$'–2' into the ground.
2. Take one of the pots and slide it over the rebar through the drain hole until it rests on the ground. This pot will be straight; in other words, flat on the ground as it would be normally placed.

(Photo courtesy of Theresa Loe)

3. Fill this first pot with soil and tamp it down so that it's filled to a couple of inches from the top edge.
4. Take another pot and slide it over the rebar through the drain hole and down to the next pot. But this second pot will be placed at an angle inside the first pot.

(Photo courtesy of Theresa Loe)

5. Fill this second pot with soil and tamp it down. Do the same until all of the pots are on the rebar post. Your Terra Cotta Tower is ready for plants!

(Photo courtesy of Theresa Loe)

Creative Repurposing 5

Thankfully, we're living in a time when this country is waking up and realizing how many things are tossed into the landfills and forgotten. Let's face it, much of what we discard doesn't actually go anywhere; it sits underground pretty much forever.

Just in case you're drawing a blank on what you can repurpose into a vertical garden, I have some great ideas to jump-start your imagination. After this chapter, I promise that you'll never look at any item that can hold soil in the same way again.

Spice Rack Planter

My husband and I were cruising around one Saturday morning taking a peek at what other people didn't find useful anymore (read: garage sales) when we came upon one of those spice racks that attach to the inside of a tall cabinet door. It looked like your typical spice rack, but I saw herbs and spices—planted. We picked it up for $3. It was a fun little unit to plant. It's a great project for someone who lives in an apartment or condominium, or anyone who has limited gardening space.

This repurposed planter is the perfect fit attached to a wall on the back deck or front porch.

Gather your materials:

1 spice rack

1 roll burlap

Scissors

Potting soil

Assemble your spice rack planter:

1. For each shelf (little basket) on the rack, cut 2 pieces of burlap in 9½" × 24" strips. Line the shelves by placing two pieces of burlap into each shelf.
2. Add potting soil to the shelf baskets (on top of the burlap).

This spice rack planter is just a jumping-off point to get your imagination into gear. Look around your house, other people's houses, and garage sales for some wonderfully unique and interesting planters!

3. Plant the shelves with herbs, strawberries, lettuce, annuals, or succulents.
4. Fold the burlap on the short sides of the shelves and tuck it into the corners of the baskets. This creates nice little pockets that keep the soil in place.
5. Water your spice rack planter gently until the soil is moist all the way through.

I have my planter on our deck, where it receives morning sun only. I think it's a great spot because this small amount of soil will otherwise dry out quickly (probably every day). At this point, I water it lightly every other day.

I fully expect this transformed planter to be seasonal and rather limited in its usefulness. Plants grown in a shallow container such as this should be a bit drought tolerant. (Herbs are a good choice.) The burlap, plants, and soil will need to be replaced each season or year. That said, it's an awesome example of recycling something into a vertical garden that would otherwise have been discarded.

Someone asked me if this was going to be enough room, even for an herb garden. My answer is yes—if you have a small family. We have a lot of people around here, so we tend to go through a lot of herbs in my kitchen. Thus, this little upright unit is constantly being harvested, which on one hand leaves plenty of room for new growth; yet I have to keep another herb garden going at the same time to keep my large family in herbs.

I can also envision sweet alyssum and baby tears spilling over the edges of these baskets, and succulents would be beautiful, too. A word of warning: containers that are this shallow will need to be watered about every other day in semi-shade and every day if they are in the sun. It isn't a big chore, but it's something to consider when you're choosing a container to repurpose.

The next time I plant this spice rack, I'll go with all herbs and strawberries and skip the lettuce. The soil ended up a bit too compacted for the lettuce's taste.

Hanging Tin Tub Garden

This is another great garden idea for vertically challenged vegetables. Jacky Alsina shared her directions with me for her hanging tin tub garden. Under the porch eaves is an excellent place for it; preferably the porch closest to the kitchen.

Look in craft stores for the inexpensive tins used for this hanging garden.
(Photo courtesy of Jacky Alsina)

Gather your materials:

3 small tin tubs

2 hooks (screw type)

Wire cutters

8' chain

6 *S* hooks

Potting soil

Hand drill with any size drill bit

Assemble your hanging tin tub garden:

1. Drill a dozen drainage holes on the bottom of each tin tub.
2. Measure the length of your tubs and mark that length on the ceiling of the porch. This is where you'll place your two screw-type hooks.
3. Using the wire cutters, cut the 8' length of chain in half. Place the *S* hooks onto the tub handles and the chain to suspend the containers.
4. Add your potting soil and plant your tin tub garden. Water small tubs like these carefully with a hand wand with an adjustable nozzle. Or better yet, use a drip system.

Upside-Down Planter

You may have seen the commercial upside-down tomato planters that have been all the rage for some years now. Topsy Turvy was the first of the upside-down planters that became available on the market and at that time, someone had gifted me one of them.

I found that the "container bag" was too small for growing tomato plants successfully, in my opinion. But I've noticed that the ones on the market today have bigger bags. In any case, purchasing one is unnecessary if you happen to have a 5-gallon bucket handy.

This project is best constructed with two people. You can't set the bucket down once the planter is filled with soil and the tomato plant is hanging out of the bottom.

Cherry tomatoes will give you the best results. Choose a plant that's quite small—in a 4" container or smaller for easy transplanting.

Create your own upside-down planter with items that you may already have around the house.

Gather your materials:

1 5-gallon bucket

1 molly hook or other strong hanger

Utility knife

Hand drill with drill bit

Scissors

1 piece of window screening or landscape fabric

Potting soil

Tomato plant

Assemble your upside-down planter:

1. You should have somewhere to hang the bucket planter immediately because it's easier to plant if it's hanging up. So attach your molly hook to the porch ceiling (or wherever the planter will be hanging) before you begin making the planter.

An alternative to hanging from a ceiling is to secure a hook into a wood beam. A 5-gallon bucket filled with soil is extremely heavy, so be warned! Also, tomato plants need 8 hours of full sun—so pick the sunniest place you can find for your planter.

2. Turn your bucket upside down and using your utility knife cut a 2" hole in the center of the bottom. Use your hand drill to make a few extra holes in the bottom for more drainage.
3. Turn your bucket right side up again. Using scissors, cut an 8"–10" diameter size circle of the screening or landscape cloth. Fold the circle in half and make a 2" slice in the middle of the screen. Continue making little cuts so that the center of the screen resembles a pizza cut into slices. Your plant is going to go through this screening and through the hole in the bucket.
4. Fill the bucket with potting soil, tamping it down to fill in air pockets. You'll want the soil line to be about 1" inch away from the lip of the bucket.
5. The best way to finish the planter is by hanging the container up at this point. This is the part where a helper comes in handy because it could take both of you to lift the bucket and place the handle over the hook.
6. Take your little tomato plant from its original container and give it a little water to make the root ball pliable. Then plant it by coming up from under the bucket and pushing the root ball through the hole and the screen. Water your planter until the water runs through the holes in the bottom.
7. At this point, I would normally say that your planter is completed. But I'm not one to ignore the fact that there's some perfectly good soil sitting at the top of the planter. I'd go ahead and plant the top of the bucket with herbs, lettuce, or a low-growing flowering plant such as sweet alyssum.

Shoe Bag Planter

You may have one of these shoe bags tucked away in a closet somewhere gathering dust. If you do, break it out, dust it off, and add some holes in the bottom for drainage (if it's lined with plastic). Fill ⅔ of the pockets with soil and add strawberries, radishes, herbs, or other small plants for a quick vertical planter that can be hung anywhere. If you don't own one, they can be found at any discount or Goodwill-type store.

Although you may never see a giant harvest from a shoe bag planter, it will earn its keep as a home for radishes, strawberries, and other small crops.

Grow Potatoes in a Garbage Bag

On traditional farms, potatoes are grown as sprawling plants. But the new trend for small-space gardeners has been to grow them up instead of out by planting them inside fencing, bamboo screens, or wooden boxes. They can also be grown in a simple garbage bag. This recycling trick is not only the cheapest way to grow potatoes vertically—but it works!

Gather your materials:

30-gallon garbage bag (use bags that are BPA free)

Soil mix—I like to use a light mix by blending $\frac{3}{4}$ potting soil with $\frac{1}{4}$ compost.

Seed potatoes—Get them ready by cutting them into several chunks (just be sure that each of the chunks has 2 "eyes" or sprouts on them). Let them sit out for several days to encourage sprouts.

Straw or dry leaves

Assemble your garbage bag potatoes:

1. Poke holes (straw size) into the bottom of your garbage bag for drainage.
2. Find an area in your yard that has full sun. Open the bag so that the sides are gathered and pressed down to the ground.
3. Add about 5" of soil mix to the bottom of the bag.
4. Plant the potato chunks so that the eyes (sprouts) are facing up about 2"–3" deep.
5. Make sure the potatoes are covered. (If you'd like to add a little more soil mix, you can do so.) Mulch (cover) with a layer of straw or dry leaves.
6. Water them well. You'll want to keep your garbage bag potatoes watered but not soaked—just damp.
7. When you see 6"–7" of leaf growth, pull the sides of the bag up a bit and add enough of the soil mix to cover most of the leaves; *leave a few leaves sticking up through the soil.* Add light straw mulch.
8. When you see the next 6"–7" of growth, repeat the process. Do this again and again as the plant grows. While the plant is growing, keep it watered but not soaking.

At some point the potato leaves will yellow and the plant will start to dry up. Stop watering the plant at that point and let it continue to dry out. Gather your potato harvest by using scissors to cut the top of the bag and slide the scissor blade down carefully. Dinner will come tumbling out!

Endless Possibilities

When you take a good look at what you're about to take to the dump or are perhaps "storing" in your garage or basement, you'll be amazed at what can be put to good use once it's transformed into a vertical garden.

Feed Bags

If you have livestock (horses, cattle, sheep, rabbits, chickens, goats, etc.) you're probably purchasing one feed or another that comes in 25-lb. or 50-lb. plastic feed bags. You may have noticed that crafters are upcycling these bags into purses and grocery totes. Well, they can also be made into hanging planters!

Feed bags have gained popularity as grocery totes … and now as hanging planters!

Gather your materials:

1 plastic livestock feed bag

Scissors

Sewing machine and thread

Potting soil

Plants

Assemble your feed bag hanging planter:

1. Sew the edges (as well as the looped tags on the back) to give them stability while they're hanging. There's no measuring involved—simply cut the bag in half (or thereabouts).
2. The stitching that's on the bottom of these bags isn't the greatest, so make a straight stitch along the bottom of the bag, following the original cotton thread.
3. From the top half (that would be discarded) cut a 3"-wide strip (the circumference of the bag) and then cut it in half so that you have two 3"-wide strips. Cut each strip in half to make the loops for hanging the planter.

4. Turn the top edge of the feed bag down and make a straight stitch around the entire top. Then take each 3" strip and turn in a small part of the edges to make a nice hem (for looks, really). Take one strip and make a loop so that the short edges come together at the bottom.
5. Turn the bag over and stitch the loop tabs into place by stitching (and backstitching) straight lines in three places on the tabs with the top of the bag sandwiched in between. The extra stitching gives the tab strength.

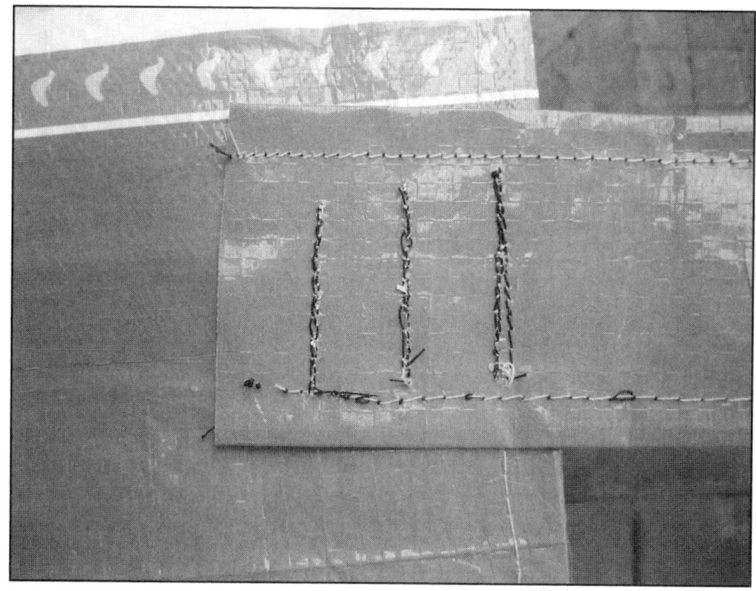

You don't have to be a seamstress to sew tab loops on the bag; just stitch them on well.

6. Cut small, vertical slices for drainage with scissors near the bottom of the bag at the back. (You can't see the slices from the front.)
7. Fill the bag with potting soil to about 1"–1½" from the top of the bag and plant.
8. Hang the bag up using zip ties.

Vegetable plants that *don't* require cage support would be best for this planter. I've planted peas in it and let them grow down as opposed to using a vine support. Strawberries, carrots, radishes, and lettuce have all been happy here as well. Experiment!

GOOD TO KNOW

If you were going to make a feed bag into a tote, you would cut the white tab off of the bottom, turn it inside out and create a rectangular bottom. Instead of tabs for hanging, you'd make long purse-type straps. Feed bag planters are just a simpler version of this craft.

Bushel Baskets

Bushel (or half-bushel) baskets are easy to find at garage sales and dollar stores. I line my baskets with plastic or landscape cloth to keep as much soil in there as I can, plus the plastic helps the soil retain moisture. You'll have to cut or poke some holes on the plastic, but you don't have to drill drainage holes into the basket.

These baskets aren't made of the sturdiest stuff, so it's best to fill and plant them where you think they're going to stay for the season. If you find that you have to move them, try to move them slowly so you don't rip up the bottom. Lifting them instead of dragging can work, but depending on the sturdiness of your particular basket, the weight of the soil can break the bottom.

Horse Troughs

Big horse troughs make excellent gardens and they're usually so large that fitting a trellis inside for vertical gardening is no problem. One thing to watch out for is that they're made of metal so they can really heat up!

Plant heat-lovers or place the troughs in a partly shaded area and grow plants that don't need full sun. Make a *lot* of holes in the bottom of the troughs using a hand drill and a drill bit made to go through metal.

Fence Panels

Have any discarded fence panels that are sitting in the side yard? Put them to good use as a climbing structure for vining plants. Just angle them against another existing fence. You can secure the panel to the upright fence if you'd like. But most of the time, you'll get away with simply letting it lean there for the season. Plant seeds along the bottom of the angled fence panel.

Tetherball or Clothesline Poles

Tetherball poles, old clothesline poles, and the like can act as the center of a "maypole" trellis. Secure several lengths of twine, jute, or string octopus-style at the top of the pole. Extend them all the way down to the ground and then tie or staple the ends to short wooden stakes that have been driven into the soil. Plant around the bottom of the twine like a teepee trellis. Because this structure uses thin supports (twine or jute), lightweight vines such as peas and beans are good plant choices.

5-Gallon Buckets

Earlier in the chapter, I detailed how to make hanging tomato planters, but honestly, 5-gallon buckets are just as useable on the ground. Just make a dozen drainage holes in the bucket, add potting soil, and plant to your heart's content. One nice thing about this type of container planting is that if you need to make adjustments for the plants, they can easily be moved into more sun or into more shade.

Ladders

Ladders offer a whole host of opportunities for gardening. The first one is to plant pole beans at the base of the legs and point the tendrils in the right direction. This is the same principal as constructing a teepee, but without the construction. If you add a climbing material such as fencing or poultry netting to the open sides of the ladder, you can take advantage of every vertical angle.

Ladders are handy additions to a vertical garden.

Some ladders (especially the versatile metal ones) have stepping rungs on both sides. If you've got this type of ladder, place runner boards that are long enough to go from one step/rung all the way through to the other side. Place the boards on all of the steps and you've got platforms for pots and other containers.

For a traditional wooden ladder with steps only on one side, screw the same number of 2" × 2" boards on the other side of the ladder (directly across from the steps); then add the runner boards. You could always leave the ladder as it is and just place terra cotta pots on all of the steps.

Discarded Table Stands

This stand was originally a table with a glass top. Completely missing the obvious potential here, the previous owners were going to toss it! But I rescued and repurposed it for my deck garden. We measured the inside of the frame where the glass was originally sitting. I didn't add the metal lip into the measurement because I knew that a plastic storage tub would also have a lip and that's what would hold the container on to the stand.

Originally headed for the dump, adding a plastic tub transformed this table stand into a raised planter.

Amazingly, we found the perfect size plastic container. We added holes in the bottom for drainage with a hand drill (and whatever drill bit was on the drill). Then we filled it with potting soil and added chives, peas, and purple sage. If I did this again I would look for a clear container (this one was light blue), as well as one that was more shallow. There's *no way* you need this much soil for this garden.

Sawhorses

If no one will miss them at your house, sawhorses are an easy way to get plants vertical. Just set two saw horses up (parallel) and place a wood plank or sheet of plywood across both horses. Place potted plants on top of the wood—plus, in a pinch it also makes a great potting table.

Old Swing Sets

A swing set frame! Oh my goodness, can you say "giant A-frame"? Use string or twine pulled taut over the top bar to create tons of climbing space. Short wooden stakes can be pounded into the ground to secure ties.

Or lay a long 2" × 4" running parallel to the top bar and secure the twine to it. Rolls of other material such as fencing or netting can be tossed over the bar as well. Anyway you look at it, swing set frames are a big plus for the vertical garden.

Mattress Springs

Mattress springs are an easy climb for veggies such as peas, beans, and cucumbers. Take a look at my friend Jenn Hammer's photo. I'll admit that the thought of a huge mattress frame sitting naked against the fence may look odd at first. But fast climbers like beans and cucumbers will quickly cover the support.

Give crib springs new life as a vertical support for vining vegetables.
(Photo courtesy of Jenn Hammer)

Broken Baby Gates and Frames

Have a broken wooden baby gate? Don't toss it! If it's jointed, hold the folded side up and spread the bottom ends out to create an A-frame. Place it in the garden and let the vining plants have their way with it. If it's flat, lean it on a wall or fence. It can be left unsecured if you're using it for the lighter vines such as peas or beans.

This is one futon that won't end up in a landfill!
(Photo courtesy of Jenn Hammer)

Other frames that are bound for the landfill may be equally handy. In the previous photo, Jenn is at it again with her futon frame!

Kiddie Pools

Hard plastic pools that have cracks in the bottom may not be fun for the kids anymore—but they *can* be fun for you! This is one of my favorite repurposing ideas, and Glenda Mills was kind enough to share what she did to create an instant raised bed with vertical structures inside.

A kiddie pool that has worn out its welcome for swimming can be repurposed as a raised garden bed. (Photo courtesy of Glenda Mills)

All you need to do is be sure there are slits or make holes in the bottom of the pool for drainage. You could then line the bottom with screening (to be sure to keep soil inside) or add a layer of small rocks to help with drainage. Glenda skipped the screen and her pool planter turned out great just using rocks.

In the previous photo you can see she has stakes inside the pools for tomatoes, but in others she had miniature teepees and cages. Even purchased brand new, these pools are incredibly inexpensive as far as raised beds go.

Garage Shelves

Free-standing plastic garage shelves are excellent for creating more space by getting containers vertical. For stability, I've placed them against house walls. The wider the shelves are, the less need for securing them to whatever they're up against.

Old Dressers

Worn-out dressers with their handy drawers are not only easily plantable, but they add character and charm to the garden. You'll want to be sure that the drawers have wood bottoms because particle board won't hold up between the soil and watering. Use dressers that have untreated wood, as well. Chemicals can leach into the soil and therefore your food crops.

Use a hand drill to make a dozen holes in the bottom of the drawers and add a layer of small rocks to help with drainage. Use lightweight potting soil, which usually means that it'll have a lot of perlite (volcanic rock that helps aerate potting soil). You can also add perlite to the soil mix.

Plastic Drums

With a little planning and sharp tools, plastic drums and garbage cans can be transformed into planters. You'll want to drill drainage holes into the bottom and possibly add some screening to keep the soil in the barrel or can. Think along the lines of a strawberry container that has the planting pockets on the sides. If you'd like more guidance, EasiestGarden.com has a step-by-step instructional DVD commercially available.

This is the 3D Barrel Garden by Easiest Garden.
(Photo courtesy of Easiest Garden)

The Basics: Soil and Seed

All gardening styles benefit from knowing the basics, and all the important stuff starts at the bottom—soil and the seeds. Part 2 is the place to learn the fundamentals of gardening. I teach you about the different types of soil, amendments, and the benefits of composting. Proper preparation of soil before, during, and after planting is the key to a bountiful harvest. I also provide detailed directions for traditional composting, as well as composting directly in a bed (the compost sandwich).

Even though your soil is prepped and ready, some of your plants may need a little extra love (and indoor shelter) before they're ready for the great outdoors, so I've included information on when and how to plant seeds for warm- versus cool-season vegetables. After learning how to start seeds and take cuttings, you may never buy a plant again!

Garden Soil 101 6

No matter what you're growing or where it ends up getting planted, successful gardening always begins and ends with the soil. Gardeners prefer to use the term *soil* as opposed to *dirt* because the latter term conjures up images of something passive and lifeless. I assure you that soil is anything but these things. In fact, it's teeming with billions of various life forms that are too small for the human eye. Organic matter, minerals, water, and air are the components of soil along with breathing organisms such as bacteria and fungi. Soil provides plants with nutrition and a stable medium for roots.

Garden soil is the lifeblood for your vegetables, fruit, and herbs. So in this chapter I talk about which type of soil should fill your garden beds and why the soil in your containers might be different. I also have directions for making your own compost—a.k.a. "garden gold"—to keep your plants happy and healthy.

Soil Medium for Containers and Raised Beds

Whether you're growing climbing vegetables such as peas in a long patio planter or nonclimbing vegetables such as lettuce in stackable containers on the porch, the medium you should be using is potting soil. Garden soil (whether bagged or from the earth), is much too dense for container plantings, which not only makes it hard for growing plant roots to penetrate and receive oxygen, but also inhibits good drainage and is extremely heavy.

Keep in mind that potting soil is often "soil-less." For example, instead of traditional soil particles, it may be made up of a combination of ground bark, peat, and vermiculite. Depending on the developer there may be a time release fertilizer sprinkled throughout or compost added. But don't be surprised if there isn't any other amendment listed on the label. This is because the soil-less mix doesn't typically have any nutritional value (unless there's compost or vermicompost added)—it's basically a medium for holding roots and water. If you're working with a potting mix, you'll eventually want to add some amendments and/or fertilizer for healthy and robust plants.

GOOD TO KNOW

Why would you purchase soil that contains vermicompost? Vermicompost (worm castings plus composted materials) contains significantly more beneficial microorganisms, enzymes, humus, and plant stimulants than regular compost. Among other things, worm castings offer these nutrients in high percentages in a slow-release form along with superior soil-binding and water-retaining abilities.

So go for the potting soil bag when planting in the smaller (the porch-type) containers—leave the "garden soil" for the raised garden beds and in-ground planting areas. Those containers that are the middle ground between containers and raised beds are the products like The Little Acre or other contained raised beds (pictured in Chapter 3). Potting soil is appropriate for these beds, but garden soil is great, too. If you choose garden soil, you can keep it on the lighter side by adding equal parts of garden soil, vermiculite, compost, and peat moss.

No matter what you're gardening in, you'll want to amend all of these soils with some extra goodies at one point or another. *Amending* is a general term that refers to improving or correcting your soil. It can be about adding organic matter such as compost for structure and stability, adding nutrients, or altering the pH balance of the soil. I talk more about amendments in Chapter 7.

GOOD TO KNOW

You can make your own soil-less blend by mixing 1 part vermiculite and 1 part perlite to 6 parts Sphagnum peat moss. Give this blend a little nutrition by adding 2 parts compost to the original recipe. Another "light" soil that isn't as heavy as garden soil but still carries some nutritional value is a mix of ¼ compost; ¼ builder's sand; and ½ peat, coir, or rice hulls. Or simply purchase a bag of potting soil and add 1 part compost to 4 parts of the potting soil.

Define "Great" Soil

What is "good" garden soil anyway? For the gardener, it's a scent and a feel. To the nose it's deep and earthy; to the hand, it's crumbly and fluffy. It's high in organic matter and rich with nutrition. The physical description is a balanced blend of organic matter, clay, and sand. This garden nirvana is referred to as "loam" or as having "good tilth."

Nearly all vegetable plants thrive in loamy soil mainly because it's loaded with organic matter (compost). With all that wiggle-room, roots penetrate easily, just enough water is retained, and excess drains easily away.

Great garden soil is easy to recognize; it has a crumbly structure much like coffee grounds.

Typically, your average backyard garden soil doesn't start out with these fine attributes (if it does, I give you full bragging rights). The soil in your yard will have the basic, general properties of the land around you, but will also have some differences, depending on where you live.

Clay Soil

You may be in a high-clay area (like me) where the soil can be described as dense (heavy) and hard to get a spade into, which means that plants will have the same problem with their roots. I actually don't mind clay soils because there's great fertility going on in there—you just have to get to it! Although the nutrition is there, the clay is locking up the nutrients that plants could be using.

The key is to break up the fine clay particles and get some aeration in there by adding compost and sand. You may have run into a common clay soil "remedy" that says to simply add sand. That's part of the solution, but not the entire recipe. If you add a bunch of sand willy-nilly to your clay, you will discover concrete. Okay, maybe not true concrete, but you're getting a pretty good image. In order to reach a state of loam, you must add the third ingredient to the recipe, which is garden compost and maybe some peat if you'd like.

Sandy Soil

On the flip side, you may have a very sandy soil, which is the polar opposite of the clay types. Sandy soil offers little in the way of nutrition for plants, and even when the gardener adds nutrients, they're quickly washed away through the sand particles. This type of soil has trouble holding water for any decent amount of time and plants dry out quickly.

The answer here is pure compost, baby. I'll tell you up front that many questions can be answered with compost as a solution. Organic matter will not only bring the necessary nutrition to sandy soils, but will also bring water-holding properties. It's hands-down one of the best things you can do for your garden or yard year-round.

The beauty of gardening vertically is that because you're not tending a lot of horizontal space, any amending that needs to be done shouldn't be expensive or labor-intensive. Remember, if you're planting in a raised garden bed or containers and purchasing bagged soil, then you won't need to perform the Basic Personality Testing in the following sidebar, as you'll be starting out with structurally balanced soil.

BASIC PERSONALITY TESTING

At first glance, it isn't always clear which type of backyard soil you've got. It may not look like a big clay brick, but it may not resemble a sandy beach, either. In fact, the U.S. Department of Agriculture (USDA) has a soil classification system that recognizes 12 basic soil textural classes. This system is put neatly together on a handy triangle diagram. Still, no matter how many times I see it, I still haven't figured out how it applies to my backyard soil, and if you've seen this diagram, you may not have, either.

So, in the spirit of simplicity and generality, gardeners have a little cheat system that'll put you on the right path to understanding your soil's personality in about 10 minutes; no horticulturist necessary.

1. Reach down and grab a handful of soil.
2. Squeeze the soil together like Play-Doh to form an oblong ball.
3. Using your thumb, gently push some of the soil forward a bit and then go back to the ball and push a bit more forward. Repeat this motion in order to create a "ribbon" of soil.
4. If your ribbon is best described as "crumbs," you've got coarse-textured soil; loamy sand.
5. If you can form a lovely, strong ribbon that just goes on and on, you have fine-textured soil—in other words, clay.
6. If your ribbon forms easily and then breaks off at about ¾", you have a clay-loam. This is medium-textured soil that's not too shabby.
7. If your ribbon reaches about ½" long and then breaks off, this is loam. You win! Cross this off your bucket list.

Okay, I'm kidding about the winning part, I'm sure your friends are excellent gardeners, too. But if you want to keep that Great Garden Soil trophy, you'll want to be sure to keep up with the great compost.

What Is pH and Why Should You Care?

Knowing a bit about pH is one of those things that might come in handy at some point and it really isn't all that hard to understand. On the most basic level, the acid and alkaline levels in the soil influence the nutrients that plants receive by making them either more available to plant roots or less available—depending on the plant species.

Chemists measure these levels by using a pH scale, which is drawn on a number line that ranges from 0 to 14. The neutral (middle) point is at 7 on the scale. Tested soils that fall below 7 have higher acidity and are referred to as "sour." Soils that test higher than 7 are more alkaline and are referred to as "sweet."

Most vegetable plants are happy as clams living in a neutral to slightly acidic soil. This is good news to those who worship compost because soils that are high in organic matter generally hover around the neutral area. By the way, another one of compost's virtues is that it acts as a pH buffer, making plants less dependent on the pH levels in the soil.

Most vegetable plants do best in a neutral to a slightly acidic pH (6–7.5). Adding compost regularly will help keep soil at a neutral place. If you find that your soil is falling over into the alkaline area, it can be lowered by adding aluminum sulfate, peat moss, conifer needles, or sawdust.

If your soil is too sour (and most plants other than those like blueberries or azaleas won't appreciate this), you can raise the pH by adding pulverized oyster shells, powdered limestone, dolomitic limestone, or sulfur. Remember, most of these pH adjusters won't change the pH overnight no matter what soil type you have. Yes, I feel compelled to thump the compost bible once again.

GOOD TO KNOW

I don't do a lot of testing and I never test my raised garden beds where I've added bagged soil and compost. For the most part, bagged garden soil falls close to the neutral or slightly acidic side, so testing is usually unnecessary. In my experience, I've found that there are more imbalances with "ground" soil than in raised beds. Even then, I don't test in-ground soil unless I feel like things are going awry. I want to make you aware of soil testing and what that's about, but don't feel that you have to test before you start planting your vertical veggies.

Testing Soil at Home

If you're interested in finding out where your soil falls on the pH scale, there are a few ways to go about soil testing, the simplest of which is an at-home soil testing kit. Let me say right off the bat that I'm not one of those people who's "against" home soil tests. There's nothing wrong with them, you just have to know that they're not going to be very specific. That's okay. You'll get a general feel about what's going on and it's kind of fun—all scientific-like.

Your local nursery or garden center will probably have soil testing kits. They look similar to pool test kits. All you do is add some of your soil, the powder that comes with the kit, and some water. Now shake it up! From there you just compare the color of the concoction with the color strips on the container. Other home tests include using simple litmus paper. This involves a soil sample, water, and a color chart to compare the dipped paper against.

This is hardly the ultimate in testing, but I've found that it's nice to know whether your soil pH is in the range that most veggies prefer (between 6–7.5) and that's enough to go on. If you have a bed intended especially for tomatoes, try to bring the pH a little lower, as they prefer their soil a little on the sour side—as do potatoes.

DIY Kitchen pH Test

This is a simple experiment you can do at home to determine if your soil falls on the sweet or sour side of things. It doesn't take the place of any in-depth testing.

What you'll need:

Measuring cup

2 mason jar–sized containers

1 cup soil sample

½ cup white vinegar

½ cup water

½ cup baking soda

1. In the first container, add ½ cup soil and ½ cup of vinegar and then mix it up. If it bubbles (or fizzes), it's alkaline. The more it bubbles; the higher the alkalinity of the soil.
2. If nothing happened with the vinegar, then put ½ cup of soil into the other container and add ½ cup of water. Mix it up and add ½ cup of baking soda. If it bubbles (or fizzes), it's acidic. Again, this test is not incredibly specific, but enlightening.

Digging Deeper: Professional Soil Analysis

Let's say that all of this science talk has piqued your interest and you'd like to know more. Or you may have moved to a new home and want to know what's in your soil (especially if you've heard rumors about chemicals being dumped nearby).

It's a simple thing to contact your Cooperative Extension office and they typically offer more precise testing for something in the neighborhood of $15. These tests are much more comprehensive and they can offer all kinds of information including what minerals and heavy metals are present in your soil, for example.

As a bonus, guidance on amendments for your particular soil is often part of the test results. You can also opt for sending samples to independent laboratories; your Cooperative Extension office should have a complete list of those labs.

Compost in the Vegetable Garden

For healthier plants, higher produce yields, and fewer garden diseases, the first place to concentrate on is the soil. The key to healthy, productive plants is soil that's rich in organic matter. Good garden soil is about having as much organic matter in the garden bed as possible. Compost is the finished product of broken-down organic plant and animal matter. In its "finished" form, compost becomes humus. Humus makes complex nutrients in the soil easily available to plant roots. Friable soil is what you get when your soil is full of life-sustaining humus, which means that it has a full, loamy texture and crumbles easily in your hands.

Compost will add any number of nutrients to the soil depending on what materials have been added to the compost pile. It adds further value by providing it with nitrogen, phosphorus, potassium, and others, plus micronutrients such as copper, iron, iodine, zinc, manganese, cobalt, boron, and molybdenum.

While humus is considered the ultimate goal, you can use compost in your garden beds at any time during the composting process. Organic matter at any stage can be used as a mulch for retaining moisture, suppressing weeds, and controlling erosion. Why not just spread on any old synthetic fertilizer into the vegetable garden, you may ask? Synthetic fertilizers may temporary mask the problem of poor soil by perking plants up with some temporary green, but they don't actually solve the problem of inferior soil. Compost, on the other hand, changes the structure of soil—making it nutritionally rich.

Obtaining Compost

There are a number of ways to get compost to your garden beds: purchase it bagged from a nursery or garden center, obtain it from your local community gardens or a municipal composting facility, or make your own.

If you buy bagged compost, your best bet is to look for a certified organic label, or better yet, keep an eye out for the Mulch & Soil Council's certification label, which lets you know for sure that the product you're buying is free of unacceptable contaminates from materials such as CCA–treated wood. Obtaining chemical and contaminate-free soil products is especially important when it comes to growing edibles. Know what you're looking for and you'll find good compost on the market.

If you have community gardens in your town, this is often a great place to collect compost. Some gardens make it available to the public and some prefer to keep it for community garden members. Many community gardens (if not all) have strict organic practices, so you can feel pretty secure that the materials going into the compost pile are chemical-free. The one thing you won't know is if someone has introduced diseased plant material into the pile. That said, this is potentially a great compost resource.

As far as acquiring compost from a municipal composting program (waste management company), I'm personally very leery about this practice. The first thing that seems like a no-brainer is that they collect grass clippings, and many people spread herbicides or pesticides on their lawns. These chemicals do not disappear, nor break down like organic materials, so you could simply be broadcasting those chemicals into your garden bed. How about your neighbor's verticillium-wilt-ridden tomato plants from last season? Those are on their way to the municipal facility, too.

If you're interested in bringing home compost from this resource, your best bet is to call and ask them what their policy is as far as what materials they accept into their program. Even then, there's no guarantee that the public necessarily follows the policy.

Making Your Own Compost

What's left? That's right; making your own compost. Creating compost is one of the easiest and least expensive ways to add nutrients and structure to your garden soil. You could toss a stick in any direction and hit information on home composting. What I find unfortunate is that much of this information (while accurate) is unnecessarily complicated. For instance, memorizing the carbon to nitrogen (C:N) ratio on every piece of material that you toss into your compost pile is not only overwhelming, it's unnecessary.

Truth be told; composting is about as simple as it gets. Anyone with a spare half hour and some organic material lying around can begin to reap its benefits. It's not rocket science, there's no magic start date on the calendar, and you don't need expensive supplies. And you don't need to do a lick of math, either. The key is to have some information, organic materials, and decide what kind of container or bin you'd like to use to put it all together. A perfectly adequate pile can be started in 30 minutes or less.

Every time I witness someone explaining C:N ratios to would-be composters, it makes me a little crazy. You can literally watch their eyes glaze over as a small part of their brain shuts down. Math has a tendency to do that to people. Here's the deal: every living thing on Earth has a carbon-to-nitrogen value. The idea behind composting is to purposely attempt to put the two together in such a way that decomposition happens effectively (and quickly) so we can put that garden gold to work.

Here's how the C:N numbers work: look at the first number (C). If that number is higher than 30, the material is considered to be a carbon product; if that first number is lower than 30, it's considered a nitrogen product.

But I promise you that the specific ratios don't matter. You don't need to memorize this stuff because most things either have more carbon and fall on the brown materials list or they're heavier in nitrogen which puts them on green materials list—and they're pretty easy to figure out at a glance. The best formula I've used is from the Master Gardeners: 1 part brown (carbon) to 1 part green (nitrogen)—just 50/50. There are a couple more ingredients you'll need, but let's meet the organic material first.

Brown Materials

The microbial and macrobial critters in your compost like carbohydrates as much as the next living thing, plus they need it for energy. It comes to them in the form of brown or carbon organic materials in your pile. Like I mentioned earlier, living things are made up of both carbon and nitrogen, but some are heavy on the carbon side and these are referred to as the "browns."

As an example, straw has a carbon-to-nitrogen ratio of about 80:1; which is 80 parts carbon (brown) to 1 part nitrogen (green). Sawdust is 500:1, and leaves are 60:1. All of these things are in the brown category.

Here are some examples of carbon materials you might find around your home:

- Dried grasses
- Cardboard (including egg cartons and toilet paper rolls)
- Paper towels
- Newspaper
- Shredded documents
- Oat hay; aged hay
- Straw
- Dried leaves (shredded)

- Chipped wood
- Sawdust
- Wood ash (not coal ash)

Green Materials

Carbon offers energy for compost critters, but it's the nitrogen that's necessary for growth and reproduction. To fulfill this requirement, you'll offer them nitrogen-rich ("green") organic materials whose carbon-to-nitrogen ratios fall lower than the number 30 on the "C" side of the equation. Examples of materials high in nitrogen are grass clippings, which are 20:1, kitchen scraps are around 15:1, and rotted manure is about 25:1. All of these things are nitrogen-rich.

Here are some examples of nitrogen materials you might find around your home:

- Green leaves and grass clippings
- Vegetable trimmings
- Green plant prunings
- Weeds (without seeds)
- Old flower bouquets and houseplants
- Alfalfa meal or green grass hay
- Used tea bags
- Coffee grounds (including filter)
- Animal manure (from herbivores such as chickens, rabbits, horses)
- Algae; kelp/seaweed
- Aquarium water (freshwater)
- Citrus peels (chopped well)

What Not to Add

So far we've talked about adding organic matter to our compost pile or bin. By definition organic matter is any material that originates from living organisms, including all animal and plant life whether still living or in any stage of decomposition. Yet there are good reasons for leaving some organic matter out of the equation.

Garden soil isn't harmful to a compost pile, but it also doesn't help it, either. People feel like it should, and get the urge to add it, but it's really not necessary. The only exception here would be the soil that was under the last batch of compost. Adding composted soil to a new compost pile would actually help inoculate it (get it revved up fast).

Diseased plants belong in the garbage can, not your compost pile. Diseases are sneaky little so-and-sos and would like nothing better than to hang around your pile just waiting to be tossed onto the next garden bed and wreak havoc. If you have plants that have been snuffed out by disease, leave them out.

Coal ashes are toxic to plants—period. They carry heavy metals like arsenic and iron along with a lot of sulfur.

Manure created by carnivorous animals is best kept out of your soil amendments. The pathogens and parasites that meat-eating animals carry are notorious for being harmful to humans. Most harmful pathogens come from carnivorous animals such as dogs and cats, which is why it's always suggested to keep them out of compost piles used for gardening. While it's true that pathogens *can* live in manures from herbaceous animals, it's much less likely. That said, the safest bet is to always compost them before using them in your garden. Dog and cat waste will also attract unwanted critters, not to mention that the smell would be a huge turn-off, so avoid composting cat litter, too.

Pesticides and insecticides defeat the purpose of your organic compost creation. So any plant materials that have been sprayed with these chemicals need to be disposed of elsewhere. The dangers of herbicides in the pile may seem obvious. After all, you're going to add it to a garden bed. But what about the pesticides? Well, pesticides are toxic to all kinds of living things (the good guys included), which will threaten the macro- and microorganisms living inside the compost pile.

Dairy products such as milk and cheese can become quite odiferous before they break all the way down. Additionally, unwanted critters like rats are attracted to this food source.

Grease and fat should be placed in a leak-proof container and disposed of instead of placed in the pile for the same reasons that you shouldn't add meat.

Four Necessary Ingredients

There are only four ingredients necessary to get you on your way to creating some life-sustaining compost: carbon, nitrogen, water, and air. You've met both the browns and greens, so where does water and air come in? Like everything, the creatures that are working away on your organic materials need air to survive.

Aerobic bacteria are the hardest-working bacteria in the compost pile, but you'll only invite them in if they have a sufficient amount of oxygen. If there's very little (or no) air in your pile, their free-loading cousins, the anaerobes, move in. Along with them comes their stinky baggage—and you don't want that. I'll get into the various ways to get air into your pile later in this chapter.

GOOD TO KNOW

Aerobic bacteria lives and occurs only in the presence of oxygen. An aerobic compost pile hosts "aerobes" (bacterial organisms) that thrive in an oxygen-rich environment. Aerobes are the most effective among the decomposing bacteria. So an aerobic pile is also referred to as an "active" pile. Anaerobic bacteria lives (is active) in the absence of oxygen. Aerobic bacterial organisms are those that occur when oxygen is absent in a compost pile.

Usually, the gardener's goal is to get their hands on some rich compost as fast as they can. One of the best ways to get on the fast track to decomposition is to have nearly equal amounts of water and air in your pile. The microbes don't function well in a dry pile, so you're looking for about 40 percent moisture, which is about as wet as a wrung-out sponge. If the pile reaches more than 60 percent moisture, it can become anaerobic; in other words, it's being starved of oxygen. The take-away here is you want equal amounts of greens and browns, keep it damp (like a wrung-out sponge), and keep it aerated.

THE PERFECT SOIL AMENDMENT

Still not convinced about the value of compost? Here are five more benefits that compost offers:

Acts as a disease suppressor. Researchers have discovered another virtue, which doesn't get as much publicity as it should: compost is valuable for plant-disease resistance. The beneficial microorganisms produced by composting organic materials render plant pathogens inactive. Potato blight, powdery mildew, and damping-off (a fungal disease) are all examples of plant diseases that compost can suppress.

Increases the growing season. Compost improves average soil structure by bringing it to a loamy, friable state. Nutritionally rich soil with good structure is able to hold heat better than poor soil. For the gardener, this means the soil warms up faster and stays warm longer, which allows the gardener to plant earlier and harvest later in the season.

Acts as a pH buffer. For most plants, the most desirable pH is neutral; neither too acidic nor too alkaline. If a gardener is generous with applying compost to garden soil, he or she doesn't have to worry about the pH levels as much—if at all (unless you have plants such as blueberries that need a more acidic pH level).

When humus is plentiful in soil, vegetable crops (and flower beds) are simply less dependent on pH levels. Due to its biochemical structure, humus acts as a buffer for soils that fall slightly to one side of acidic or alkaline. This not only takes the guesswork out of an average pH level, but in many cases, it can take the pH factor out of the equation entirely.

Saves water. Compost increases soil's capacity to hold water by a wide margin. For instance, a dry soil low in nutrients may only hold 20 percent of its weight in water. Comparatively, a dry soil that's high in organic content can hold up to 200 percent of its weight in water.

Reduces water runoff. Due to the poor crumb structure of soil that's low in organic matter, it can be washed away easily by storms or even everyday watering. Lost topsoil results in even lower fertility, creating a vicious cycle. But compost preserves and enhances soil structure and helps fight erosion, keeping healthy soil under the plants where it belongs.

Composting is sustainability at its finest for plants and the other living organisms on this planet. In the vegetable garden, it provides structure, adds nutritional value, suppresses disease, increases the growing season, saves money, and reduces water runoff. All of these benefits allow the plants to work at maximum capacity, and that means more vegetable bounty for you! Compost is garden soil's best friend.

How Important Is a Compost Bin?

Truthfully? Not very important at all. In gardening circles the term "bin" is used loosely. For instance, three pallets that are standing on end and connected together surrounding a compost pile is referred to as a bin. A compost bin can also be made from a large piece of galvanized wire or hardware cloth formed into a circle, wire fencing around steel fence posts, stacked cinderblocks, or picket fencing. All are just barriers that keep the pile rather pulled together, but aren't a bin in the sense of a completely enclosed container with a lid. The idea is to confine the materials so that there's a combination of green materials, brown materials, moisture, and air. And for that, you don't need a bin at all; the compost pile could even be on the open ground.

GOOD TO KNOW

The soil around the bottom of a compost pile is rich and ready for planting. You can take advantage of that by planting a vining crop such as mini pumpkins, climbing cucumbers, or nasturtium flowers around the base of the compost pile. These plants will soon cover the entire bin with attractive vines, flowers, and fruit.

Stylish market compost bins aren't essential, but they aren't without their merits. For one, you may be worried that dogs, raccoons, and other critters will be encouraged to loiter around, given all the free food. This is a legitimate concern and it's the best way to be sure that things placed into the compost stay put. Containers with secure lids are a great idea, especially if you're adding kitchen scraps to your pile, which is really what the critters are after.

Another reason is aesthetics. Some people just prefer the "look" of a compost bin to the look of an open pile, which may seem messy. They may feel that it looks better to their neighbors, as well. An enclosed bin retains moisture longer, so that can be a plus.

My point isn't to discourage you from purchasing a compost bin; rather, I want to encourage you to compost even if you don't have a bin or would rather not spend money on one—because it really isn't a necessity for creating compost.

Simple Hoop Compost Bin

This do-it-yourself basic hoop-style bin goes together in a hurry, but it's easier with two people.

What you'll need:

Work gloves

Safety goggles

12½' of hardware cloth or horse fencing

Pliers

Heavy wire snips

Metal file (optional)

4 steel fence posts (also called *T* posts)

Post slammer or sledgehammer

4 metal clips or wire

Assemble your compost bin:

1. Roll out the hardware cloth or horse fencing. The fencing will have sharp, exposed wire ends. Using the pliers, bend those ends back onto themselves. Hardware cloth can be cut at a cross wire to prevent loose wires from poking out. You can use the file for smoothing out any sharp edges, as well.
2. Place the steel posts at evenly spaced intervals inside the hoop. The posts should be pressed taut against the wire hoop.
3. Pound the steel posts into the ground with the post slammer or sledgehammer. Some people find that their hardware cloth is so heavy they forego the steel posts entirely. Hoops made of heavy wire without the post support will work just fine, they're just not as stable. Another advantage to making the hoop without posts is that when it's time to turn the pile, the hoop can be lifted off and placed next to the compost pile. To turn, just fork the pile into the bin again.
4. Secure fencing to each post with clips or wire.

A hoop bin made of wire or hardware cloth makes a perfect compost bin.

Compost Pile Maintenance

If you have a 50/50 pile of greens and browns, you have half of the ingredients to make compost. As I've mentioned, there are four ingredients necessary in order for your organic material pile to break down: equal amounts of greens and browns, plus water and oxygen. Decomposition will be at its speediest if the piles are built touching the bare earth. This is because naturally occurring micro- and macroorganisms will be quickly drawn to your pile, which is exactly what it needs to get things cooking.

There are several ways to keep air in a pile, but for simplicity's sake, the easiest way is to turn it. You could use a pitchfork and turn the pile over in the spot that it's in or move the pile from one place to an area right next to it. It should be turned once or twice a week—or never (more on that in the next section), depending on how fast you'd like it to break down.

Your pile should stay about as damp as a wrung-out sponge. To keep moisture in my piles, I water it while I'm turning or moving the pile. I add water before I begin turning and add more after I've turned a couple of forkfuls, as well as after it's all been turned.

Don't be fooled into thinking that if you stand there with the hose running onto it while you sip on a Mojito, that the pile will become damp all the way through. These piles get dense and I assure you, the water won't reach all the way into the center or bottom of the pile.

Remember that we don't want the pile sopping wet, just uniformly moist. This means that during heavy winter rains, you may need to place a tarp or another cover over it. Conversely, if you want to keep the pile from drying out under the summer sun, you may need the cover (or tarp) to trap the moisture.

Composting Styles: Hot, Warm, or Cold

No matter how much time you have (or the length of your attention span), there's a composting style for everybody. A "hot" or "fast" compost pile is exactly what it sounds like: core temperatures can range from 113°F to 160°F. A "warm" or "hybrid" compost pile doesn't get as hot, nor break down as quickly, but it's respectable and requires very little attention. Then again, if you just want to toss it and forget about it, a "cold" or "passive" compost pile might be for you—although this method takes the longest amount of time for the organic matter to decompose. (I have to admit that I gave those compost piles that are "halfway" tended the moniker "hybrid" or "warm," because I had no idea how else to describe them. This is the composting technique that I use most of the time.)

How you balance the brown and green materials will determine whether you have a hot, warm, or cold compost pile.

Hot (fast) composting tips:

- A hot compost pile needs at least 3' × 3' × 3' in order to heat up, and ideally even a bit wider and taller. Between 3' and 6' tall and 6' deep is a good rule of thumb. Hot pile temperatures range anywhere from 113°F to 160°F, and you can have compost in about 8 weeks.
- Try to build a hot pile all at once. Once you have a nice-sized pile going, don't add any more to that one and just work it until it's broken down (finished) into fluffy, earthy humus. Periodically adding more materials to the pile will add more decomposing time.
- Turn (aerate) this pile a couple of times per week to get things hot in there. Don't try to rush it any faster by turning every day; the bacteria have to have some time to do their thing.
- Keep the moisture steady. Get water into the pile as you're turning or moving it. I like to have a garden hose handy and I water things down after I've moved a few forkfuls of organic matter. Remember, not sopping wet, just thoroughly damp—like a wrung-out sponge.

Warm (hybrid) composting tips:

- The warm compost pile is a happy-medium pile that's neither ignored nor pampered. Start it out as if you were building a hot pile. Add a lot of green and brown materials in equal parts. Then wet it thoroughly and toss it around making a great mix.
- Turn the warm pile once every 2 weeks. At that time add some water, as well. It gets just enough love to keep things smelling nice and moving forward, but it might take 8 to 14 weeks to break down.

Cold (passive) composting tips:

- A cold compost pile breaks down slowly at temps that are 90°F or lower. At these kick-back temperatures, you'll have garden gold in about 6 months to a year or more. It'll depend upon the season and what you're tossing in there. Still, who cares? Where's the fire, right?
- This is the no-work composting system. Toss your greens and browns together, willy-nilly, add some water once in a while (or never), aerate now and again (or never), and Mother Nature is still going to do what she's great at: decompose the material.
- A cold pile is one that you may want to keep hidden; because it does take so long to break down, you may consider it unattractive. There's also the chance that this type of pile could potentially get smelly—especially if you never aerate (turn) it. It can take a year to collect compost from a cold pile.
- Cold piles have a certain desirable quality, however. High temps in a hot compost pile kill off certain fungi and bacteria that help suppress soil-borne diseases in the vegetable garden. These beneficial microbes are left intact in the humus produced by a cold pile.

GOOD TO KNOW

Homemade compost doesn't always look exactly like the stuff purchased in bags. You can improve its appearance and leave the materials that haven't completely broken down behind by sifting the final product. Remove the bottom of a wooden crate with handles, and use a staple gun to attach wire mesh to the bottom of the crate. Put some compost in your new sifter and sift it to perfection!

Composting Myth-Busters

There are some uninformed souls out there giving composting a bad rap, and they're typically false accusations. When people cite reasons for not composting, they often roll out one of the following composting myths. Fortunately, most of these statements are either blatantly false or have just a ring of truth to them. Still, most issues are easily resolved.

Myth #1: Compost Piles Smell Bad

In the years I've been composting, I've never had a bad-smelling compost pile. At least not for long. You almost have to try to create a bad-smelling pile. I promise that it doesn't take much to keep it as nature intended—fresh like the forest floor.

Organic matter that's composting has a woodsy smell. But it's entirely possible to put ingredients together in a way that encourages bad bacteria to move in and give it a bad odor. Even then, the problem can be easily remedied by adding some carbon such as paper or leaves, or by simply turning the pile over.

Odoriferous piles are usually the result of too much nitrogen (for example, a thick layer of grass clippings) or the addition of meat or dairy waste. The solution would be to add carbon materials and turn the pile over, or in the case of meat and dairy, remove it entirely. Another reason for unpleasant odors is if the pile is getting sopping wet. It can become anaerobic, which is again about adding some browns and aerating the pile.

Myth #2: Composting Is Complicated

After all, creating good soil is all about measuring and science, right? Wrong. As we talked about earlier, you don't have to do a single calculation to create gorgeous soil for your garden; just remember the 50/50 rule.

Myth #3: Composting Is Expensive

The only part of composting that could be costly is if you run out and buy an expensive composting bin. Yes, the pricey bins are out there, but the good news is that you don't have to purchase any kind of compost bin. It can be done on the bare earth, out in the open or within the confines of wooden pallets or other materials.

Myth #4: Composting Takes a Lot of Time and Effort

You have to be a true work horse to keep a compost pile hot and moving. Wrong again! I'm a fairly lazy composter. I certainly don't work hard at it, and I have a ton if it. I gather the goods necessary, pile it up, wet it, and turn it once in a while. Nature does the rest. (I swear!)

Myth #5: You Need a Lot of Land to Compost Correctly

Another rumor that doesn't hold water. Urban and suburban gardeners can compost in a 3' × 3' × 3' space. People who live in apartments and condominiums are perfectly capable of composting, too. They might use a small tumbling bin, a plain garbage can, or a vermicomposting (worm bin) system.

Myth #6: You Need to Add Special Compost Activators or Starters

While it can be nice to toss in a commercial activator to get things going, in no way is doing so a necessity. In fact, I've never used one in my piles, and I have some fast-producing piles. You can also toss some rabbit food (alfalfa pellets) or some compost from a previous pile as a free activator. In fact, if you start a compost pile in the same place that you had the previous one, that's the best "activator" of all!

Myth #7: A Compost Pile Attracts Animals

Okay, I'm going to give you this one, they *can* become attracted. If you add dairy products, meat, fat, and other food waste, then rats, raccoons, dogs, and stray cats will almost certainly make their way to your pile and dig around looking for some Friday night takeout. Sometimes raccoons or even the family dog will even go for veggie or fruit scraps in a compost pile. This is why I suggest that gardeners only add kitchen scraps to enclosed compost bins such as the plastic ones with four sides and a secure lid. An open pile should be loaded only with yard wastes and nonfood items.

Make Your Bed with a Compost Sandwich

A compost sandwich is my favorite way to create a new vegetable garden bed that's ready for next season's crops. A compost sandwich isn't worked the same way that a traditional pile is worked; it's utilized a bit differently. Usually, a compost pile is worked and turned until the materials are broken down to a level that's now considered humus. At this point, the compost is added to a garden bed or as a top dressing around plants.

A compost sandwich is made in layers and is meant to be left exactly the way it was placed—it's never turned or worked. In other words, it's left to decompose just as it sits, while creating a new garden bed by default. I've made compost sandwiches in areas of my yard where there previously was no bed; such as at the end of my lawn.

There are many advantages to starting a garden bed by making a compost sandwich. It'll have very few weed problems, and if they do appear every so often, they'll pull easily out of the crumbly soil. These beds offer excellent water-holding capabilities, thus making terrific use of any rainfall. Best of all, it'll be pliable, nutritional, and ready for your spring seedlings!

I usually start this bed in the late summer or early fall to give the materials time to compost for planting the following spring. If you decide to build this sandwich in the spring, be sure to add topsoil into the layers and maybe some peat moss for good measure. You could go ahead and plant it with veggies immediately while everything is breaking down. The plants would still do great—but next year's crop will be fabulous.

What you'll need:

Cardboard

Newspaper

An assortment of carbon materials (browns) such as leaves, straw, weed-less grass hay, newspaper, shredded bark, 100 percent cotton clothing, etc.

An assortment of nitrogen materials (greens) such as grass clippings, vegetable peelings, seedless weeds, perennial plant clippings, coffee grounds, tea bags, etc.

Topsoil

Manure from herbivores (chickens, rabbits, horses, etc.—no dog or cat poop)

Water source and hose

Make the compost sandwich:

1. Cover the entire garden area with cardboard. It can be corrugated cardboard or whatever you have. Then water the cardboard down. You'll be watering between each layer to get everything moving along down the decomposition path. You're not trying to flood it, but the sandwich needs to be wet.
2. Lay newspaper over the cardboard. You'll want to make this layer about 2" thick. Then water it all down.

3. Spread a layer of greens over the newspaper. If you choose grass clippings, keep the layer thinner than the other materials as the grass tends to compact and not let air inside. Add water.
4. Spread a manure layer over the greens, and then add a thin layer of topsoil. At this point, you'll go back to your carbons and this time you might use straw instead of newspaper (this will depend on what you're using for browns). You can also go back to newspaper. Add water.
5. The last layer will be topsoil; add water. Now—other than watering the sandwich if you have dry weather—leave it alone. Don't do a darned thing to it all winter. You're going to be so thrilled with the soil in your new bed next spring.

Compost sandwiches are the simple way to create a new garden bed ready for next season.

I can't stress enough that composting of any kind is an art—not rocket science. While there's certainly a basic chemistry to it, you don't need to measure and get precise. Make your compost sandwich the best you can and use varying materials while creating.

A Plant Primer

So when should you plant your vertical vegetable garden? Where do you get the plants? Should you begin with seeds, *starts,* or cuttings? What are *open-pollinated, heirloom,* and *hybrid* seeds? Can you save seeds from the plants you've grown and replant them? In this chapter, I answer all of these questions and offer guidance so you can decide what works best for you.

In almost every case, vegetables can be both started from seed and purchased as baby plants (starts). But there are some sound reasons to choose one or the other, depending on the plant.

Some Like It Hot: Warm-Season Vegetables

Warm-season vegetables are those that you plant and grow throughout the late spring and summer months. They're the sun-worshippers that thrive in the summer heat. They need outdoor temperatures to be 60°F or higher and the soil at 50°F or higher for seed germination.

The middle-to-late spring through the beginning of summer are the right times to plant. Before planting, find out how many days each particular variety needs in order to mature and provide fruit, or you may find that you've planted too late to enjoy a decent harvest.

GOOD TO KNOW

Don't forget to take a look at how many days of warm temperatures you have in your area. Search online for the American Horticultural Society (AHS) Heat Zone Map at www.ahs.org/pdfs/05_heat_map.pdf. You'll be on your way to a successful harvest if you coordinate your days to the number of days a vegetable variety needs in order to mature and produce fruit. If you're not sure, contact the Cooperative Extension Office in your county for guidance.

Once the last frost has come and gone in your area (and soils are 50°F or higher), you're free to get these heat lovers out into the garden. However, to get a jump on the season, start your crops indoors under lights 4 to 6 weeks before that last frost rolls around. We often start long-season crops this way because they need many warm days in order to mature. Eggplants, melons, peppers, tomatoes, and summer and winter squash are good candidates for starting early indoors.

While most vegetables fall into the warm-season camp or the cool-season camp, some will cross over these categories depending on when they're started in the garden and if you're manipulating conditions by using your microclimates.

It's the fruit of the warm-weather crops that we're after on most of these plants. The obvious exception would be the New Zealand spinach, which is considered an alternate to the more common cool-weather spinach for the summer.

Warm-season crops include:

- Bush beans
- Cabbage
- Cantaloupes
- Cauliflower
- Celery
- Corn
- Cucumbers
- Eggplants
- Melons
- Mustard
- New Zealand spinach
- Okra
- Peppers
- Pole beans
- Pumpkins
- Summer squash
- Sweet potatoes
- Tomatillos
- Tomatoes
- Watermelons
- Winter squash
- Zucchini

Many of these warm-season vegetables listed, such as cabbage and onions, may not be suitable for vertical gardening (although those varieties can be grown at the base of a vertical garden). But I wanted to add them here to give you as many examples of warm-season vegetables as possible.

If you take another look, you'll notice that I have both summer and winter squashes on the list. It may look like a mistake, but the term "winter squash" is rather misleading, as they're *grown* as a warm-weather crop. The winter squash moniker refers to the fact that these squash varieties are kept on hand in cellars everywhere as food throughout the winter months. We'll talk more about winter squash varieties in Chapter 10.

Some Like It Cold: Cool-Season Vegetables

Cool-season vegetables can be grown at either end of the warm months; both late winter/early spring and fall. Soil temperatures should be 45°F to 55°F for seed germination, and outdoor temperatures need to be about 40°F to 60°F.

When I first started growing cool-weather crops, I experimented with the plants to see which "cool end" they preferred the most: spring or fall. I was living in the San Francisco Bay area, and when I planted my broccoli and cilantro plants in the spring, the quick temperature rise as the season went from spring to summer caused them to bolt quickly. From then on, I planted my broccoli and cilantro in the fall, and found that they produced a better harvest for longer. I suggest you try planting at each end of the season, as well, to see which works best for you.

Some cool-season crops can go well beyond cool and into the freezing cold—all the way through snowy winters. You'll need a little help from your handy-dandy hoop house and cold frame (see Chapter 1), as well as some mulch (see Chapter 8).

On the following cool-weather vegetable list, you'll typically find that it's the leaves, stems, flower buds, and roots of these plants that we enjoy in the kitchen—the exception being fava (broad) beans and peas. We're actually after the fruit of these cool-weather legumes.

By the way, there are several "cross-over" cool-season vegetables (e.g. carrots, beets, Swiss chard, cabbage, potatoes, and leafy greens) that can be harvested in the summer as long as they got a good start in the cool weather.

Cool-season crops include:

- Asparagus
- Beets
- Brussels sprouts
- Cabbage
- Carrots
- Cauliflower
- Cilantro
- Endive
- Kale
- Kohlrabies
- Leeks
- Lettuce
- Parsnips
- Peas
- Radicchios
- Radishes
- Rhubarb
- Spinach
- Swiss chard
- Turnips

As I mentioned in the last section, not all of these vegetables are suitable for vertical gardening. I included them for clarity.

Why You Might Love Starts

Starts are baby plants that have been grown from seed (by someone else) and are ready to go into the garden bed. You'll find them in nurseries and garden centers often in six packs or 4" plastic pots. There are several good reasons to use starts in your garden plan:

- You're new to gardening and would like to skip the long road that's starting from seed.
- You'd like to get your garden in immediately.
- You're interested in obtaining varieties that you know will thrive in your area. Local nurseries tend to bring in plants that do well in their area.

I adore starts and it's not because I don't adore seeds. The majority of my vegetables are grown from seed, but starts are simply a leg up to the season. Plus, they're instant gratification when you've been working hard to create a garden bed or vertical gardening system.

While you're shopping for vegetable plants, take a look around. Are they being housed under an awning or shade cloth? Or are they out in the sun? If these are sun-loving plants and they've had a barrier between them and the sun, then you'll need to harden them off before planting them into a sun-drenched garden bed. When you bring these little guys home, place them in the sun for an hour and then bring them back under cover. Lengthen the amount of time they spend in the sun for a week or so and then plant them into their permanent bed.

By purchasing starts instead of starting from seed, you don't have as many variety choices. That being said, if you're a new gardener, plant starts can give you the confidence to plant a garden in the first place.

DOWNER

The downside to beginning with plant starts is that they have outgrown their containers by the time you get them. It may not be obvious from the top, so slide a plant out of its container to see if the roots have all but sucked up the space in there. Lots of roots circling the bottom is a sign that the plant is already stressed and growing incorrectly. Leave this poor plant at the garden center. You're looking for a plant with lots of healthy, white roots that still have wiggle room in the soil and container.

Propagation: Free Plants Forever

It's quite possible to never purchase another plant again. Ever. I admit that this isn't the norm because we gardeners are easily seduced by new-to-us varieties and this is as it should be. The truth is that you don't have to rely on anyone but yourself and your mother plants in order to grow gardens forever, if you use propagation techniques. Propagating is simply making more plants from mature plant tissues (sometimes called the mother plant). There are many ways to go about propagating plants, including seeds, cuttings, division, leaves, roots, layering, bulbs, grafting, etc. However, for the purposes of this book, we're going to discuss the most basic practices: seeds and softwood cuttings. Vegetables, herbs, and cane berries are typically propagated in these ways.

For me, propagation has always been one of the best parts about growing things. Exploring ways that plants reproduce and practicing and perfecting the techniques used to do so have been both exciting and satisfying. The simple act of saving seeds for next season's garden is my favorite propagation method. I have to say, though, that taking cuttings is just as easy.

MAKE MORE STRAWBERRY PLANTS

One of the easiest ways to propagate strawberry plants is by their runners (stolons), which is a stem that grows horizontally against the ground from the base of the original strawberry plant. This shoot produces a "plantlet" with roots at the end (nodes). Almost all of the strawberry plants you purchase will grow runners, whether they're June-bearing, everbearing, or day-neutral varieties.

The roots at the end of the runner eventually touch the soil and begin to grow into the ground, which, in effect, creates a young strawberry plant (plantlet) exactly like its mother. You can take advantage of this situation by placing a small container filled with loamy, nutrient-rich soil underneath the runner's roots. Bendable stolons make directing them a simple task.

Once the runner is sitting on the soil in your container, place a staple-shaped wire over the middle of the runner to hold it in place to give the plantlet roots a chance to take hold. After several weeks or when you notice that the plantlet is well-rooted, simply snip the runner away from the mother plant. You now have a brand-new plant to start another strawberry bed, plant in a container, or give to a friend.

If you hadn't cut the runner from the mother plant and it was left to its own devices, the runner would eventually shrivel up and die as the little plantlet takes hold. This would expand the original strawberry bed naturally.

Starting with Seed

I have to admit that I'm a seed freak. I can't think of anything better than collecting seeds from my friends and my own plants. When I get them from my favorite seed companies, the gorgeous plant art that covers the packets is just a bonus.

One great reason to start your plants by seed is, of course, the variety. No nursery or garden center can compete with the vast selection of varieties, colors, flavors, and characteristics offered in seed catalogs. Armed with several of these—in print or online—the world is your oyster.

Growing from seed is also the cheapest way to get plants without sacrificing quality, and the inexpensive/good quality combo always gets my attention. On top of that you also get *more* plants at the same time, which is an additional savings.

If you're interested in starting out with strong, healthy plants, seeds are the way to go. As the grower, you have the ultimate in quality control. There's no telling how long seedlings have been sitting around a garden center becoming root-bound and jostled about. If you start some of your seeds directly into the garden bed, you'll avoid transplant shock entirely. Starting from seed not only gives you a jump on the season, it allows you as an organic gardener the assurance that your plants (and vegetables) have been raised chemical-free.

Some people don't like to start from seed because they've never done so before and it can seem difficult. For the most part, it really isn't. I'll admit that occasionally seeds won't germinate, a soil fungus will wipe them out, or the cat will dig into their little growing pots and mush them. It happens. But usually, most of them germinate and a couple die—and that doesn't matter because you suddenly realize you have no idea where you're going to put all of these seedlings once they've grown up. The seeds may germinate faster or slower than you thought, but growing plants from seeds is exciting and satisfying.

STARTING YOUR VEGETABLE SEEDS INDOORS

Following is a list of the items you'll need to start your vegetables indoors to get a jump on the season. This is a tried-and-true guideline that works for most gardeners, but feel free to experiment!

Indoor seed-starting materials list:

Seeds. Seeds are pretty easy to come by; you can purchase them, ask friends for them, or have saved them from the previous year. If they weren't harvested from last season (fresh for this year), then you can check for good viability (meaning to see if they'll germinate). Dampen a paper towel, sprinkle some seeds over it, and fold the towel into quarters. Place it into a plastic baggie and set it somewhere that's warm or on a sunny windowsill. Within about 7 to 10 days, the seeds will begin to germinate—or not. If over half of them have sprouted, this is decent viability. If it's much less than that, then you may want to start with fresh seed.

Seed starter mix. The medium that is used to germinate seeds is a soil-less mix. It's usually peat moss, vermiculite, perlite, or coir, depending on which company mixes it. In any case the advantages are the same: good drainage, lightweight, and no surprise diseases.

Containers. There are many seed-starting containers to choose from; you just have to decide which you like best. Some kits have 6-cell packs that come complete with a bottom tray to catch water. Some come with a lid or dome to keep the soil mix moist until the seeds sprout. There are also little Jiffy pellets that expand with water and peat pots in all sizes. Clean yogurt containers, egg cartons, and toilet paper rolls will perform the same way and won't cost you a nickel.

Labels. Chances are you'll never remember which tomato variety you planted in which container, so be sure to label every single cell. If you've started your seeds in one of those large plastic containers where the individual cells are attached, you can get away with labeling just the first row of seedlings for a short time. But when you transplant them into larger pots, the "memory game" won't be nearly as fun as it was when you played it as a kid.

Heat. There's a lot of flexibility here; you can get away with a warm room and heat from the overhead lights for many seeds. I've had great luck starting tomatoes with no bottom heat, for instance. But peppers are another story; they really, *really* love a warm bed. Depending on what type of plants you're starting, you may or may not need bottom heat such as coils or a heat mat.

Light. You're going to need a light source after the seedlings show up, so you might as well think about it now. Grow lights are special lights that contain the full color spectrum. They're also expensive and unnecessary for starting vegetable seeds indoors. If the seedlings you're starting are going to be transplanted and live the rest of their lives outside, you can skip the specific grow lights. Instead, pick up some ordinary fluorescent shop lights and fixtures and hang them over the grow trays (very close to the top of the plants). If you hang them from chains, you can move them up little by little as the plants grow. Can you put them in a sunny window? Sure, but my experiences with this technique have always ended in leggy (long, lanky, and weak), wimpy plants.

Water. You want good humidity to surround the seeds before they germinate, so keep the soil mix damp and perhaps covered with a plastic lid or something similar. Once they've popped up, remove the lid and water sparingly. Don't let it dry out … but over-watering can do just as much damage.

Indoors or Outdoors?

Technically, all seeds can be started indoors or outdoors. Whether you choose one or the other will depend upon which vegetables you're growing, your climate, and soil temperatures. The reason you start some vegetables indoors and under lights before they can safely be planted outside is twofold.

The first reason is that some vegetables (or a vegetable variety) need a high number of days to mature before they'll produce harvestable fruit, which is very often the case with tomato varieties. If you waited until the soil was warm enough to plant them directly into the garden bed, you may not have enough warm days to reach the maturity date of that plant before the temperatures dropped and killed the tomatoes.

Now let's say that you do have enough warm days for the plant to mature. You still may opt to start them indoors because you'll be harvesting the fruit earlier in the season if you get them growing early. In the case of pepper plants, it's best to start them indoors (early) because they are unforgiving when it comes to cold soils. In fact, they take their time coming up inside a warm house if there isn't any heat underneath them. Don't get me wrong, without the bottom heat, they *will* come up—eventually.

On the other hand, cool-loving plants like broccoli and cauliflower are often started indoors because although they need the cooler temperatures while actively growing, their *seeds* won't germinate if the ground is too cold. Plus, depending on how fast your zone warms up, you may need all the cool growing days you can get in order for the heads to form properly. Giving them a head-start before the last frost date puts them on the road to success. Starting broccoli indoors in late winter/early spring will give you a harvest before the heat comes and convinces the plant to bolt (and it doesn't need much convincing).

Vegetables that benefit from an indoor head-start include:

- Broccoli
- Brussels sprouts
- Cabbage
- Cauliflower
- Celery
- Cucumbers
- Eggplants
- Melons
- Peppers
- Pumpkins
- Onions
- Tomatoes
- Zucchini

Cool-weather *root* crops don't like their roots disturbed, and since they enjoy the cool earth anyway, they're usually planted directly into the garden. These include:

- Beans
- Beets
- Carrots
- Parsnips
- Peas
- Radishes
- Turnips

Leafy vegetables such as lettuce and Swiss chard can be planted directly, but can be started indoors, as well—say going into the fall and winter/early spring.

Another reason that I like to start many vegetables indoors is that I feel like it gives them a leg-up on pest defense. They feel more "substantial" and prepared to take on an earwig or two as a young, leafed-out plant as opposed to a spindly seedling.

Hybrid Seeds

Hybrid vegetable plants (and seeds) are the offspring of two plants that are of different varieties, as produced through human manipulation for specific genetic characteristics. These varieties have a closely related gene pool (basically recycled), which leaves very little genetic diversity in the plant.

What this means to you is that if you save the seeds from a hybrid vegetable, the resulting plant will not resemble its parents. Conversely, it will end up with throwback genes and could look (and taste) like any number of things hiding inside its genetic code. In other words, hybrids don't breed true, and some hybrid seeds are sterile, so they won't geminate at all. Hybrids are created and often owned (patented) by the company that created the cross, as well.

GOOD TO KNOW

An F1 hybrid (first filial) is the first generation of plants created by crossing two different plant varieties or types. To produce consistent results, the same cross has to be made each year.

From a commercial point of view, hybrids make sense as they do what they were bred to do, which is to provide the following: fruit with uniform color and shape, high yields, fast production, longer shelf life, and the ability to withstand long truck hauls. But a little may have been lost in the translation: excellent flavor, variety of flavor, genetic diversity, many colors, and heritage.

In no way am I implying that they shouldn't be grown. There are some wonderful hybrid vegetables that are quite flavorful and productive. I'm suggesting that we celebrate and enjoy what open-pollinated and heirloom plants can offer us as gardeners and food growers; and we recognize that perhaps vegetable gardens shouldn't be predominately hybrid varieties.

Please don't confuse hybrid plants with genetically modified (GM) or genetically modified organism (GMO) foods. They're not the same thing, although they're often used interchangeably—and incorrectly. GM foods are crop plants that have been genetically engineered or modified to have specific desirable traits. In laboratories, plant geneticists isolate a single gene from one plant and insert it into a different plant to create one that's drought-resistant or herbicide-resistant. Genes from nonplants such as bacteria and fish are often inserted into crops, as well.

Open-Pollinated Seeds

Open-pollinated plants (and seeds) are those that have been pollinated naturally by bees or other insects, mammals, or the wind. Seeds created in this way will "breed true" to their parent plant. In other words, if you plant the seeds from open-pollinated plants, they will grow up to look, act, and taste like the plant they came from.

If you're interested in saving seeds from your vegetables to replant next year, open-pollinated plants are the right ones for the job. Neither open-pollinated nor heirloom vegetables are owned by anyone—they belong to everybody, so feel free to save and grow to your heart's desire.

DOWNER

If you're going to save seeds from your vegetables, you should know that varieties as well as family members are capable of cross-pollination with one another; some more readily than others. In order to be certain that you're getting pure seed from a specific variety, techniques such as bagging, caging, or distance may be used. Some gardeners simply save their seeds and don't worry about keeping a specific variety pure.

Heirloom Seeds

First of all, heirloom (or heritage) plants are always open-pollinated as they're a subset of the open-pollinated category. What makes a vegetable (or flower or fruit) an "heirloom" plant? This depends on whom you ask.

Some gardeners feel that the variety has to have a generational history—and a story—to be called an heirloom. Purists don't label an open-pollinated plant an heirloom unless it can be traced back 100 years. Although there's no official standard for heirlooms, today most gardeners agree that an heirloom variety is an open-pollinated variety that's been handed down through families for 50 years or more.

Heirloom plants were brought to America by immigrants worldwide. They're rich with culture, and many have wonderful stories attached to them. As favorite family vegetables whose seeds had been saved and passed down from generation to generation, they're worth a second (and third) look.

You may be wondering *So what's the big deal about heirlooms?* or *Why should I plant them in my vertical vegetable garden? Give me one good reason.* I'll give you five *great* ones.

Fantastic Flavor

Any vine-ripe vegetable grown in the home garden (hybrids included) beats store-bought vegetables any day of the week. Still, most heirlooms have a flavor factor that the prolific hybrids simply can't match. Many of these food plants were selected and handed down through generations expressly for taste. Commercial hybrids are created for uniformity in color, shape, size, yield, transporting abilities, and the ease of machine-harvesting. This isn't to say that there are not delicious hybrids—there certainly are. But heirlooms come in many varieties (and therefore, flavors) to please the palate.

Adaptability

Heirloom plants have an inherent ability to adapt naturally to their environment. This includes acclimating to the soil they're planted in as well as the specific climate. Historically, as vegetable varieties adjusted to their environments, they also developed resistances to local pests and diseases. In other words, there were strong, viable plants suited to every area. Because these plants evolved naturally, nothing was mechanically altered (and therefore given up), and their fruit was able to retain their delicious flavors.

Control

Food is a basic human necessity, and the person who controls the seed controls the food supply. Unfortunately, a handful of companies control all of our commercial seeds worldwide. This is possible because they literally own the seed. Heirlooms are owned by no one—and everyone. They give you control over how your food is grown, what's put on it, and which vegetable types you'll grow.

Links to Our Heritage

Heirlooms weren't always called "heirlooms." In fact, the term *heirloom* wasn't even used until the 1980s! They were simply traditional vegetables grown in gardens everywhere. They were the staples of life. Some of the heirlooms that have been preserved by family seed-saving go as far back as 2,000 years or more. Connected to those seeds is the history of our ancestors and who they were, which gives us a basic definition of who we are. In a nutshell, seeds are a living heritage for people. We can hand down antique furniture, jewelry, and paintings, but none of these are living things.

Along with great stories come the variety names. Monikers such as Beaver Dam, Moon & Stars, and Lady Godiva pumpkins, Rattlesnake and Dragon Tongue beans, Tall Telephone peas, and Mascara, Drunken Woman, and Frizzy-Headed lettuce always bring a smile … and excellent questions. By the way, these characteristics can also belong to other open-pollinated vegetables that aren't necessarily considered heirlooms. They just may not have the extensive history of their counterparts.

Genetic Diversity

This trait is what allows so many varieties of vegetables to exist in so many different areas. Gardeners in Alaska can have potatoes just like gardeners in California—all because there are varieties adapted for each environment. Diverse genes are also what protect us from having one pest or disease attack a crop like beans and wipe every one of them out. This is a true drawback for those planting a monoculture.

Perhaps one of the best examples of a monoculture gone wrong is the Irish potato famine of 1845. During the 1840s potatoes were Ireland's main food staple. Every farmer grew a variety called Lumper, which was vegetatively cloned; each plant was genetically identical. During the fall of 1845, North America inadvertently introduced a deadly fungus that attacked the potatoes, destroying every single potato crop. Thousands of people fled Ireland in an attempt to survive, and more than a million people died of starvation due to the lack of genetic diversity among their plantings.

GOOD TO KNOW

One of the simplest things you can do for your garden is to make it impossible for any insect or disease to devastate a plant species. The best way to achieve this goal is to practice broad genetic diversity by planting different vegetable varieties.

Hardening Off Seedlings

"Hardening off" bridges the gap between coddling your baby seedlings indoors (where they were completely protected from the weather, good or bad) and immediately starting life outdoors in their permanent garden bed or container. It makes no difference that the seeds you started are sun-loving tomatoes. If you toss them outside willy-nilly you'll have burned leaves, which could mean dead plants, considering that seedlings don't have very many leaves at this point.

It's not just the bright sun that we're worried about, either. Temperatures inside your home (and under lights) compared to the outdoors (especially at night) can be enough to kill young plants if they aren't acclimated first.

Bringing your little ones outdoors under a protected (covered) area for a couple of hours starts the 7-to-14-day hardening off process. Notice that I said *under cover,* as opposed to in the sunshine. I prefer to get the seedlings outside to feel the breezes and the difference of the degrees before I pop them out into the sun.

I bring them back indoors after 2 to 3 hours. After a couple of days vacationing in a shaded area, I place them out in the light for 1 hour—and it'll usually be morning sun because it's gentler on them than the brilliant afternoon sun. I always bring them back into the house at night, although after their daily dose of sun, I may pull them back under cover and not bring them into the house until evening. Day after day, I increase the time that they spend in the sun. After about a week, I let them spend the night outdoors in a covered or protected area.

The stretches in the sun become longer until about 2 weeks after I began the process, I simply plant them into their new home. So, what's with the 7-to-14-day process? Well, sometimes when I'm hardening off seedlings that will live the rest of their lives receiving only morning sun, I never actually make it to the 2 weeks because the plants simply don't require it. As you experiment, you may find that you don't need the full 2 weeks with certain plants, either.

Cuttings

Vegetative propagation includes techniques that involve separating some vegetative part (roots, shoots, and leaves) of a plant in order to create new ones, as opposed to collecting seeds, which is sexual propagation (reproduction).

Cuttings are a favorite propagation technique because some plants can be difficult to start from seed. Germination may not be a problem, but the plant may have been cross-pollinated by a different variety, and the plants grown from the seed may not end up like the parent plant at all. Taking cuttings gives you a clone of the mother plant.

Cuttings can be taken as softwood, semi-hardwood, or hardwood, and this is determined by the type of plant. Taking softwood cuttings (or slips) from plants such as blueberries, kiwis, cane berries, and herbs is one of the easiest forms of vegetative propagation. Depending on the species, after your cuttings have been started, you'll have new plants in 2 to 5 weeks!

Gather your materials:

2"–3" long cuttings from the tips of the plant (young foliage)

4" clean containers (usually plastic pots)

Sandy potting soil; dampened thoroughly (enough to fill the containers)

1 pencil

1 plastic bag (that fits over the top of your containers)

Popsicle sticks or long twigs

Scotch tape

Scissors or sharp knife

Rooting hormone (optional)

Gloves (optional, but necessary if using rooting hormone)

Propagate plants from cuttings:

1. Prepare the new plants' temporary home by filling the 4" containers with the dampened potting soil.
2. Snip cuttings off of an older plant just below a node, which is the leaf "joint" where they attach to the plant stem. You'll want two or three nodes on each cutting and a pair of leaves should be left at the top of each one.
3. Take the cuttings and clip off any flowers or buds. Then clip off the leaves at the bottom of the cutting so that the leaf node is left exposed. There will probably be just a single pair of leaves at the top.
4. Make a hole in the potting soil with the pencil so that you don't have to shove the soft stem into the soil.
5. There are two ways to go about this step. The first is to put on gloves, dip the bottom of the cutting (including the nodes) into a powdered or liquid hormone, and then place it into the potting soil. This gives the cutting a greater chance at successful rooting.

   All you'll need is plain tap water for the second way. Many cuttings will grow roots even without using the hormone. For that matter, many plant species will root in plain water. So, this part is up to you. The 4" container can start more than 1 cutting at a time, so feel free to place three or four in the pot.

DOWNER

Rooting hormone can be a gardener's best friend, but it's also a hazardous material. Please read the manufacturer's directions before opening the container and always use gloves when handling the hormone. Keep out of the reach of children.

6. Put the cuttings into the soil and press the soil up around them. Make sure the places where you cut off the lower leaves (node) are buried.
7. Place a plastic bag over the entire container and secure it around the container edge with tape to keep the cuttings moist while they grow roots. To prevent the bag from touching the cuttings, you may want to use a couple of Popsicle sticks or small branches (that are a little taller than the slips) and stick them into the soil before covering with the plastic.
8. Keep them in a warm place, but shade them from any direct sun. When you see new growth on the cuttings, make some slices with scissors into the plastic as air vents for a few days. If the soil begins to dry out, gently add a little water using a small watering can. Remove the bag completely after several days of having vents.
9. When you see more growth, use a spoon to scoop under the cuttings and place each little plant into its own container.

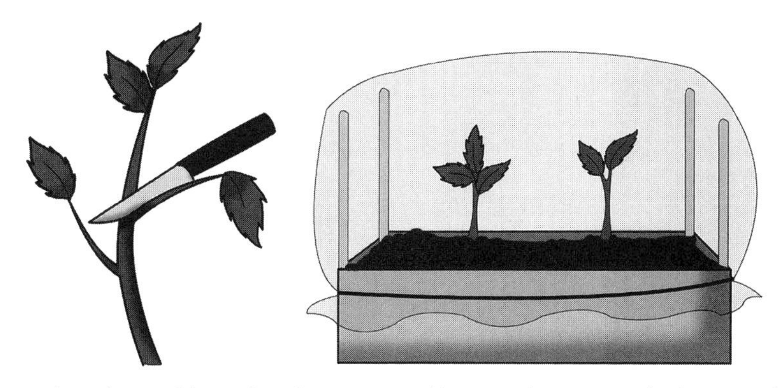

Propagating plants with cuttings is an easy and inexpensive way to obtain more plants.

WILLOW WATER AS A NATURAL ROOTING HORMONE

Instead of using a synthetic powdered product on the end of cuttings, you can make your own natural rooting hormone from the growing tips of a willow tree. When willow branches are soaked, both indolebutyric acid and salicylic acid leach into the water from the branches. Indolebutyric acid is a hormone that stimulates root growth, and there are high amounts of it found at the growing tips of willows. Salicylic acid triggers the plant's defenses, which will help protect the new cuttings from pathogens.

"Willow water" can stimulate root growth in your plant cuttings without requiring a trip to the nursery or the use of potent chemicals.

Gather your materials:

4–5 or more willow branches

1 pitcher or large bowl for soaking willow branches

Pruners

1 vase or water glass for soaking plant cuttings

Container with soil

Make willow water:

1. Cut several short branches from a willow tree; the more you take, the stronger the rooting properties in the water will be. Branch pieces don't need to be a particular size, just as long as they fit into your water container (below).
2. Fill your pitcher or large bowl with boiling water and place the branches into the water. Let the branches soak for 24 hours.
3. Fill the vase or water glass with the willow water, leaving the branches behind in the larger container.
4. Place the plant cuttings that you want to root into the vase or water glass and let them soak up the willow water overnight.
5. The next day, add the cuttings to the container of soil that you prepared to house them in while they're growing roots.

Now that you're familiar with different propagation methods, it's time to reap what you've sown and focus on what it takes to grow your vertical garden.

Tending the Vertical Vegetable Garden

3

Now that you've planted your seeds, it's time to focus on the day-to-day workings of your vertical vegetable garden. Feeding and watering are clearly important for successful gardening, but there are a lot of questions that can come with it. In Part 3, I answer these questions with details on the essentials of plant and soil nurturing. Look here for information on mulching, organic amendments, crop rotation, and pruning.

By now, you might have noticed that you're not alone in your garden, so I also include information on all of those creepy-crawlies, how to tell if they're friend or foe, and why you should know the difference. When handling the foes, organic and least-toxic pest control is my preferred method; however, if you *have* to break out the big guns, I explain how to handle chemical products safely.

Feed, Water, and Nurture 8

The truth is that plants only need soil, sun, and water to survive. But we not only want them to survive, we want them to *thrive* and produce, and do so with as little trouble for us as horticulturally possible. While I can't promise zero work on the gardener's end, I can suggest some techniques that have proven successful for many vertical gardeners.

From watering and feeding to rotating and pruning, the nurturing advice I give in this chapter helps guide you in your vertical vegetable gardening endeavor.

Small-Time Irrigation

Water is at the top of the list when it comes to what's important to a garden. In order to save water, time, and money and have healthy plants with higher yields, you need to get the water to the roots directly and steadily. Unlike traditional horizontal plots, your vertical garden is smaller and planted more intensely, and this leaves you with a wide variety of choices.

How much and how often should you water? I wish I could offer you a definitive answer, but there are several things to take into consideration. Would you describe your soil as loam, sand, or clay? Do you live in a windy area? Are these new transplants, seeds, or established plants? Raised beds or containers? Because every garden is different, there are some general rules-of-thumb that you can tweak according to your specific growing needs.

Seeds. For successful seed germination the top 2" of your seedbeds should be kept evenly moist. This means that you'll be watering (or misting) them every day, especially if they're outdoors where they'll dry out faster.

Seedlings or new transplants. Both of these are in a transitional phase of their lives and need to have evenly moist soil for a while. Continue watering the seedlings every day (like seeds), and after 2 to 3 weeks graduate to three times a week. Transplants should get the same treatment because their roots are becoming established in their new home. Remember that anything you transplant should be done so in the cool morning hours and watered before and after it's planted.

Established herbs, shrubs, or trees. If the plants in question have been recently planted, they'll need regular watering until they become established. For herbaceous perennials this might mean 6 months, and for trees or shrubs a year or more. Again, how many *times* they're actually watered depends on your soil (clay, loam, or sand) and the particular plants. I keep everything evenly watered if it's new to my garden, and as woody plants and trees become established, I water them one to two times a week during the hot months.

GOOD TO KNOW

When discussing woody plants and trees, deep watering less often is almost always preferred over shallow watering more often. Deep watering encourages deeper roots.

Vegetable plants. Most vegetables have shallow and medium roots, which means their roots grow anywhere from 6" to 24" into the soil. This list includes beans, beets, broccoli, cabbage, carrots, celery, cucumbers, eggplant, kale, leeks, lettuce, mustard greens, onions, peas, peppers, potatoes, radishes, spinach, and summer squash. Since the top several inches of soil dry out the fastest, these plants will need more watering, for less time.

Deep-rooted vegetables whose roots are capable of penetrating past 24" include asparagus, parsnips, pumpkin, rhubarb, sweet potatoes, tomatoes, watermelon, and winter squash. These plants need to be watered less frequently, but deeply.

As far as my vertical vegetable and herb gardens go, I tend to keep the vegetable's soil damp and the herb's soil on the drier side. It's a simple, no-frills strategy that works for me. If this is your first garden and you worry that the soil looks wet on top, but aren't sure that you watered for long enough, try this:

1. First water your garden as usual.
2. Two hours later, use a hand trowel to dig down into the bed (avoiding plant roots) and see where the damp soil ends. If the root zone of those plants isn't damp, then water again. Damp soil should be dark and moist like coffee grounds or chocolate cake. If the soil is soggy, water less next time.

Sprinklers

Sprinkler systems are almost always the least efficient way to water any garden. First of all, sprinkler heads are fairly inaccurate. I realize that they can be positioned to spray water into a general direction, but in reality, they end up watering places that simply don't need water, such as concrete patios. They also tend to wet plant foliage, which can encourage fungus to take up residence on the plants. Second, they allow for a ton of evaporation, which isn't efficient or cost-effective.

On the other hand, if all you have on hand is a sprinkler, it'll still get the job done. To make it the best that it can be you'll want to have the right sprinkler head for the garden; for instance, don't have a full circle head placed between the garden bed and the driveway. Also, adjust the heads so they don't aim the water at leaves (or walls), and use a timer to prevent over-watering and simplify the task for you at the same time. One final trick with using sprinklers is to turn the water pressure on low. Doing so prevents water runoff by allowing the water to sink slowly into the soil.

Hand Watering

Hand watering may be a time-consuming task, but there's something about it, right? It's so relaxing and mindless that it's almost zen-like. Maybe it's just me. Anyway, if you have a very small garden or are gardening on your patio or deck, hand watering may be the only thing you need for irrigation.

There are all types of gadgets for hand watering, such as long watering wands and simple to fancy-schmancy water nozzles with 15 different settings to help you get the job done. Don't even get me started on the vast array of watering cans on the market; from colorful plastic to hand-painted, galvanized metal, they can become somewhat of an addiction (or so I've heard).

Watering by hand puts you in a good position to control the water flow, too, which is awesome because this is another way that gives the soil a chance to absorb water and avoid wasteful runoff. Just think *slow.* I recently heard this technique described as watering as you would pour tea: carefully and slowly. The idea is to linger as you water the bed or container so that it's absorbed into the soil before adding more.

If you're hitting plant leaves with water instead of just the root zone, you could encourage leaf diseases just like the sprinklers can. Your best bet is to be sure to water early in the morning so that any leaves you do hit have a chance to dry off during the day. Also, just like the sprinkler, if you're watering soil areas where there aren't any plant roots, then you're probably watering weed seeds. So do your best to keep the water under the plants.

Soaker Hoses

Soaker hoses look like garden hoses with pores. When attached to a hose or spigot they slowly "sweat" moisture into the soil. The water seeps slowly along the entire length of the hose directly at the root zone. You can place them down along a row of vegetables, weave them throughout a garden bed, or make a circle at the drip line of a fruit tree. They're not a good choice for containers but are great almost anywhere else.

Drip Irrigation

Drip irrigation is truly the best bet for most gardens. Drip systems do a bang-up job delivering the water to all the right places. They're great for vegetables gardens, fruit trees, and berry cane beds. In containers is where they truly shine. We have some hot summers here in California, and it isn't unusual to water containers twice a day in the summer (depending). Drip irrigation makes my life much easier when it comes to my containers.

Here's how it works: it's a tube system that has evenly spaced drip emitters coming off of the tubes. Usually, each plant has its own personal emitter, although when plants are closely spaced, they may share one. Some systems come with the emitters factory installed into the tubing, and some are made so that you attach them yourself.

Another benefit to using a drip system is that the entire garden surface area isn't soaked, so you've got fewer weeds popping up. Drip systems are often sold in kits, but can be purchased piecemeal, too. Admittedly, these systems take a bit more planning, but they aren't difficult to assemble. The drawback to using drip emitters is that they may periodically become clogged, so it's important to check each emitter from time to time and watch your plants for signs that they're lacking water.

Rain Barrels

I think rain barrels are one of the best ideas ever. Not that it's a modern thing by any means; people have been collecting rainwater for centuries, but it's once again coming into vogue. Water falls from the sky seasonally every year, and collecting and storing some for the sunny days ahead is brilliant. Think about this: off of a 1,000-square-foot roof, 600 gallons of water can be collected from only 1" of rainfall. That's a lot of free water.

A rain barrel is a large container (barrel or not) that's connected to the end of the downspout that runs along the roof on the outside of your home. As the rain washes off the roof, instead of going through the downspout and becoming stormwater runoff, it's collected in a barrel. Barrels have a screen over any open areas to prevent mosquitoes from reproducing and to keep debris out of the container, too. Most barrels have a spigot at the bottom so you can fill a watering can or bucket.

There's some controversy on collecting rainwater this way. Some people claim it is "natural soft water" that doesn't contain chlorine, fluoride, minerals, or any other chemicals. Others claim there are impurities in the water due to the fact that the roof itself could leach chemicals or that air quality may be poor (pollution), therefore tainting the rainwater. One more interesting thing about collecting rainwater—it may be illegal. That's right, apparently some states believe the water falling over your home doesn't belong to you, and therefore you cannot harvest it. So, double-check that before you install one.

HEAT STRESS IN PLANTS

When the summer temperatures crank up, we aren't the only ones feeling the heat. High temperatures can be a challenge for plants, and leaf tissues will show signs of heat stress, along with blossoms or fruit. Learn to recognize the signs of heat stress in your plants:

- Foliage and flowers wilt
- Foliage becomes sunburned
- Leaves rapidly turn brown and fall off
- Vegetables drop fruit prematurely
- Vegetables become sunburned

The permanent wilting point (PWP) is the point in which there's the minimal amount of moisture in the soil before the plant wilts. Ideally, you'd like to keep the plant hydrated above this point so that the plant doesn't wilt at all—and thus, isn't forced to recover. Of course, if the plant is too far past the PWP, it reaches the point of no return. Here are some ways to keep the moisture in the soil and available to plant roots:

Add compost (organic matter). Add it to your garden beds whenever you can. Compost (among other great things) adds tilth to the soil and allows soil to easily retain moisture.

Mulch! A few inches of mulch in your garden beds and landscape will go a long way in retaining moisture.

Water in the early morning. There's less evaporation in the cool of the morning and it allows the foliage to dry before heading into the evening (which can encourage fungus).

Water plants deeply. If you're watering by hand, set the nozzle to a slow setting so the water has a chance to seep down into the soil and get to the bottom of the roots.

Use drip irrigation. This is perhaps the best bet for irrigation, as it delivers water directly to plant roots without wasting a drop of water.

Set timers. Timers are inexpensive and can be attached to almost any type of watering system. Timers set you free, save time and money, and provide a steady water supply for your plant. Once you use them you'll wonder how you ever got along without them.

Use water-holding gels. Containers dry out faster than any raised garden bed. Check out the water-holding gels available on the market; they're a lifesaver, especially if you're away for a few days in the summer.

Mulch Matters

A multitude of problems can be avoided by mulching your garden. Mulching is the second best thing you can do for your vegetable garden aside from composting (and compost can be used as a mulch, by the way; see Chapter 6 for more on compost). The praises of covering bare soil in landscaping, flower borders, and vegetable beds is sung by gardeners everywhere—and for good reason. Mulching saves you time, money, and effort. Adding just 1" to 2" of mulch helps to …

- **Retain water.** Mulch shields the soil from the sun and drying winds, therefore cutting down on evaporation. Mulched vegetables need less watering time.
- **Control weeds.** Weed seeds find it difficult to germinate when the light is blocked out by mulch. Even when a weed does start to sprout, mulch will often smother it before it gets a chance to grow up.
- **Prevent erosion.** Mulch hugs organic matter (compost) and amendments close to the ground so plants aren't washed away by water and rain.
- **Protect roots.** Consistent soil temperatures provided by mulch protect plant roots and keep them cool for the summer.
- **Act as a disease barrier.** Mulch, when added seasonally, adds a protective layer between plant leaves and last year's fallen leaves or debris. This added layer is important because as water hits the ground, it normally splashes back up and hits plant leaves, potentially transferring bacteria to new growth.
- **Condition soil.** Organic mulch eventually breaks down or composts, thereby building the soil. All life forms in the soil appreciate this benefit.

Numerous materials can be used as a mulch, and offer the same benefits no matter what type of garden or yard. For example, bark, rocks, and gravel are common inorganic mulches found under trees or in foundation landscaping. However, vegetable gardens are for the most part annual plants (there are exceptions, some herbs for instance) and the bed is often disturbed by harvesting, turning, and replanting seasonally and sometimes more often.

Organic Mulch

Mulch for vegetable gardens is usually an organic material that breaks down quickly. My favorite organic mulches are newspaper, straw, cardboard, shredded leaves, and grass clippings. By the time the next season rolls around, these materials will have broken down and become part of the soil; no fuss, no muss.

Newspaper is my hands-down favorite material to use in the garden. Most people have it on hand—or their neighbor does. It hangs around long enough to suppress weeds and hold in moisture, yet thoroughly breaks down much to the delight of all the soil critters. No matter which mulch you choose, keep it 1" to 2" away from the stems of vegetable plants, as that practice could end up inviting fungal diseases and pests.

I start out by placing four or five layers of newspaper around the plants in the bed and then dampen the papers to hold them in place while I add a small layer of compost or soil over the top to hide the evidence. I may have to do this twice; once at the beginning of spring and once in the middle of summer. It's a cheap, easy, and earth-friendly technique that works!

Cardboard works as well as newspaper, but it'll take a little longer to break down. Also, a thick layer of compost has the same mulch-type benefits of any other organic mulch.

Inorganic Mulch

Inorganic materials such as landscape cloth or black plastic can be an extra-effective mulch for hot soil lovers such as pumpkins, watermelons, and cucumbers. Black plastic is especially nice because not only does it smother weeds and retain moisture, but it also ramps up the heat in the soil.

If you choose to use solid plastic, you'll want to add holes (or slits) so that water can flow freely underneath. Some landscape fabrics don't allow water to penetrate fast enough, so be sure to check that and add holes if necessary. Both materials are easily removed at season's end.

Mediterranean herbs appreciate mason's sand as a mulch, which not only helps drainage, but reflects precious sunshine that many herbs crave. If you would like to do this for your herbs, remember that you don't want to come home with "play" sand (for sandboxes) or "beach" sand (from beaches). Their grains are too fine for this job. If your herbs are in vertical containers, add about ½" to each pot; in beds, add 1". You'll find mason's sand at any home improvement center.

NPK: Nitrogen, Phosphorus, and Potassium

If you've been anywhere near bags of fertilizer, you're sure to have noticed three big numbers that look something like this: 10-20-10. Those numbers refer to the weight percentages of the three "main" nutrients that plants need to thrive. Plants need a different amount of each of these nutrients at different times in their development. Here's what the bag is telling you:

"N": The first number represents the percentage of nitrogen. A bag with the numbers 10-20-10 is made up of 10 percent nitrogen. Think of it as steroids for leaf and stem development. A nitrogen boost gets things growing in a hurry and is always appropriate for leafy vegetables like lettuce,

Swiss chard, and kale. It makes sense because it's the leaves of these plants that we're interested in eating. On the flip side, too much nitrogen for too long can give you the wrong results. For example, tomato plants enjoy a nitrogen boost while they're young, before they develop flowers. Too much nitrogen after flowering, however, results in a lot of foliage and not so much fruit. That's clearly no *bueno*.

"P": The second number stands for phosphorus. So the 10-20-10 fertilizer is made up of 20 percent phosphorus, which is a valuable nutrient for the development of flowers and fruit. It also encourages good root development. A phosphorus boost is appropriate when crops such as tomatoes and pumpkins are beginning to flower. Like everything else, it's best to offer a fertilizer with a higher level of phosphorus during this time, but don't overdo it. A plant's ability to take up necessary micronutrients is reduced if there's an overload of phosphorus in the soil. That said, phosphorus is hard to come by in natural surroundings, so it's difficult for the plants themselves to obtain too much.

Aside from both the positive and negative effects on plants, there's also an impact on the surrounding environment. Phosphorus can cause pollution in groundwater, rivers, lakes, and streams as rain easily washes it from the soil and into nearby waterways. Local aquatic and wildlife are also affected.

"K": The third number in this equation is potassium or potash. A 10-20-10 fertilizer contains 10 percent potassium, which works in tandem with phosphorus for fruit development and healthy root building. Potassium also helps with photosynthesis as well as protecting plants from disease.

It's highly unlikely that your plants will get too much potassium as only small amounts are available naturally. What makes it even more difficult is the fact that potassium fixes itself to the heavy particles of clay soil, making it even harder for plants to get it. We've established that these major nutrients are important to plants, but there are plenty of other micronutrients that are just as important. Although there's nothing wrong with adding fertilizers to your garden (in fact, your plants may need them), you don't necessarily have to rely on the fertilizer industry for the right mix. There are other ways to get what your plants need by way of organic amendments, as you'll see next.

Organic Soil Amendments

At any nursery or garden center, you can find ready-made fertilizers or amendments. Some are organic mixes, some not. In any case, it's certainly the most expensive way to get nutrition to your plants, and you don't always need all the ingredients that come in the bag. Consider the following specific organic amendments instead.

General organic amendments:

- **Compost.** As I've stressed throughout the book, organic matter matters. It adds structure, tilth, and micronutrients that your plants need. Use it.
- **Aged animal manure.** "Aged" means that it's been sitting around in a pile somewhere for at least 6 months or more. Depending on the animal, you run the risk of burning your plants if you toss it on fresh from the beast. By "animal manure," I'm speaking of herbivores (plant-eating animals) such as horses, cows, sheep, goats, rabbits, chickens, alpacas, and llamas. Do not use manure from carnivores (meat-eating animals), such as dogs and cats. They contain harmful pathogens that you don't want in your garden.
- **Pelleted lime.** The main reason to add pelleted lime to garden soil is to balance out a soil that's too acidic. Remember if a soil is too far on the acidic side or too far on the alkaline side, plants have a hard time absorbing nutrients. Lime can help you out if you have a low pH. But lime also adds calcium, which is important, as it aids in absorbing nitrogen and building strong cell walls.

Nitrogen sources:

- **Grass clippings.** Green grass clippings fresh from the lawn are a terrific source of nitrogen. Of course, they should come from a lawn that's chemical-free.
- **Worm castings.** Worm castings and vermicompost are excellent soil conditioners. The difference between the two is that the pure castings are separated from the castings and the composted materials in the worm bin. I have no idea how they can separate the two, but use whatever you can get.
- **Alfalfa meal.** Alfalfa meal isn't only a great source for nitrogen, but it brings along other trace minerals, too. I also use it in my compost as an organism activator for soil.
- **Blood meal.** This is exactly what you think it is: dried and powdered animal blood. It's loaded with nitrogen but be sure to bury it in the soil because it can attract critters.
- **Fish meal or emulsion.** When I use this source I prefer the meal because I'm told that the emulsion washes away faster. You're correct in thinking that this is going to smell for a bit. But your plants are going to worship you.

Phosphorus sources:

- **Bone meal.** This slow-release source is a powder made from animal bones that is a good source of calcium and phosphorus. It's great for encouraging good root growth and flowers.

- **Rock phosphate.** The good news about rock phosphate is that it breaks down very slowly. The bad news is that it breaks down very slowly. Rock phosphate doesn't become fully available to plants for up to a year. If you're not in a hurry, get it into the soil because it'll release phosphorus for the long haul.
- **Soft-rock phosphate (or colloidal phosphate).** This is a phosphate made of clay particles surrounded by natural phosphate. It's another slow-release amendment that becomes available to plants during the second year after it's added. It brings micronutrients to the soil, as well.

Potassium sources:

- **Greensand.** Greensand is a good general soil amendment that's also high in potassium. It's mined from mineral deposits that come from the ocean floor.
- **Granite meal.** Granite meal builds structure and improves soil drainage. It's a finely ground granite rock that releases potassium—slowly.
- **Kelp meal or liquid seaweed.** This is an excellent fertilizer that contains trace minerals and hormones that help create strong roots. It's another garden gift from under the sea. Kelp meal is derived from dried seaweed. Kelp meal and fish emulsion is a popular fertilizer combination.

Compost and Manure "Teas"

Now we've come to a currently hot gardening and scientific debate. Does "tea" brewed from compost or aged manure really give plants (soil) a burst of life-supporting nutrition? In my experience, the answer is yes. I feel that my plants have benefited from applications of both compost and manure teas throughout the growing season. Am I a scientist? No. But I can tell you that I'm not alone in my theory, and many gardeners have used this simple organic brew as fertilizer for their garden with what they've referred to as "impressive results."

There are those in the industry who claim that this just ain't so, because the fact is that it hasn't been scientifically proven. Fair enough. Maybe the only way to decide is to try it for yourself and see if it makes a difference for your plants. One thing is certain; it won't do any harm.

If you'd like to give it a go, you'll need:

- A burlap sack, cheesecloth, or fine netting in which to strain the compost
- A 5-gallon bucket or any large, watertight container
- Water
- Compost

You're after a ratio of about 1 part compost to 5 parts water. (You can use more compost to brew a stronger tea.) Add the compost to the burlap sack or tie it into the netting like a giant tea bag.

Fill the bucket with water and place the tea bag into the bucket to steep for 24 hours to several days. Remove the tea bag from the bucket, empty the contents into your garden, and use the tea to water your plants. I like to water with the tea every couple of weeks.

You can also make manure tea by replacing the compost with aged manure and using the same method as above. The key here is *aged* (6 months or more)—not fresh from the local horse stable. Also remember that the manure should be from herbivores such as cows, horses, goats, sheep, llamas, alpacas, or my personal favorite—rabbits.

Of course, you can always take the easy route and purchase prepackaged organic moo poo tea that comes in "personal-sized" tea bags, which are perfect for small gardens. Check Appendix B on where to get my favorite teas made by Authentic Haven Brand.

Using prepackaged moo poo tea bags by Authentic Haven Brand are the easiest way to brew up a batch of manure tea.
(Photo courtesy of Annie Haven)

About Crop Rotation

Crop rotation is a fundamental organic gardening practice. It's not a hard concept, but it does take a little bit of planning—mostly in the way of pencil and paper to keep track of plants. The good news is that many of us are keeping track of the plants and where we plant them anyway, so it may be no extra effort for you to give it a try.

It's all about changing the beds that crops are grown in from season to season; usually on a 4-year cycle. Rotating crops season after season is a great idea for a couple of reasons. First, this technique preserves biological diversity, which can prevent the buildup of diseases in the soil, as well as pests that attack specific crops. Second, crop rotation can prevent the heavy feeders from depleting the soil of nutrients; in some cases, actually *improving* soil fertility.

I'll be the first to admit that crop rotation can be challenging in the smaller beds (or containers) that are often used when gardening vertically. If this is the case for you, don't worry; it's not a deal-breaker by any means. You may not have enough beds or what-have-you for this practice, but it's a good technique to know and you can use it in any way you see fit.

There are a couple of ways to go about crop rotation. The first is about sorting by individual vegetable families, the family rotation plan; and the second is based on the nutritional needs of these plants, the soil fertility rotation plan.

Family Rotation Plan

Each vegetable plant falls into an extended family (see the following sidebar). These families share specific enemies (pests and disease) that enjoy feasting on them. For instance, fungal diseases such as blight attack potatoes and tomatoes, which are both in the nightshade family. Flea beetles like these crops, too. Brassicas such as broccoli and kale belong to the cabbage family, and are pestered by cabbage looper moths and cutworms. You get the idea.

Let's say that you have four beds and the first one is planted with lettuce, kale, and broccoli. While these vegetables are growing, they're potentially calling in pests and diseases that are specific to those crops. Both the pests and the diseases can be harbored in the soil just waiting for the same type of plants to come along again. So the next time you plant these vegetables, you'll plant them in the second bed, effectively tricking all the undesirables.

Do the same thing with the remaining beds—just move each plant group to the next bed. Incidentally, if you only have three beds rotate those three or rotate some of the vegetables out and into containers. It doesn't have to be perfect or done according to some printed plan. It's just more food for thought (pun intended). This is probably the most common type of plant rotation.

ALL IN THE FAMILY

Nightshade Family (*Solanaceae*)

- Eggplant
- Peppers
- Potatoes
- Tomatillos
- Tomatoes

Gourd Family (*Cucurbitaceae*)

- Cucumbers
- Melons
- Pumpkins
- Squash
- Watermelon

Pea Family (*Fabaceae*)

- Beans
- Fava beans
- Peas
- Soybeans (edamame)

Carrot Family (*Apiaceae*)

- Carrots
- Celery
- Cilantro
- Dill
- Fennel
- Parsley
- Parsnips

Beet Family (*Chenopodiaceae* or *Amaranthaceae*)

- Beets
- Quinoa
- Spinach
- Swiss chard

Mustard Family (*Brassicaceae*)

- Arugula
- Asian greens (bok choy)
- Broccoli
- Brussels sprouts
- Cabbage
- Collard greens
- Kale
- Kohlrabi
- Mustard greens
- Radishes
- Turnips

Onion Family (*Alliacaeae*)

- Chives
- Garlic
- Leeks
- Onions
- Shallots

Sunflower Family (*Asteraceae*)

- Endive
- Lettuce
- Radicchio

Soil Fertility Rotation Plan

There's another way to rotate crops, and that's according to whether they are leaf, root, flower, or fruit crops. This system is not necessarily better or worse than rotating families, but it's a good plan for small gardens (the situation for many vertical gardens). Plus, it's actually easier to keep track because the groups seem more obvious. This style is based on rotating crops that have different nutritional demands.

As you learned earlier in the chapter, the three major players in soil nutrition are nitrogen, phosphorus, and potassium (NPK). Vegetables use different amounts of these primary nutrients and it's easy to see which ones if we look at the part of that plant that we're eating.

It's important to remember that although each crop type needs more of a certain nutrient, they *all* need each one of these nutrients on some level. I don't mean to imply that the major nutrient these plants crave is the *only* one they need to thrive and produce.

Leaf crops. As the name implies, these plants are grown for their leaves. Or at least their aboveground foliage parts, since in the case of broccoli and cauliflower we're actually eating unopened flowers. The leaf group includes leafy greens and those in the cabbage family, such as broccoli, cabbage, herbs, kale, kohlrabi, lettuce, mustard greens, spinach, and Swiss chard. Leafy crops need a lot of nitrogen in the soil to produce the lush leaves that we're harvesting.

Root crops. This group is all about the part we eat that grows underground, such as beets, carrots, garlic, onions, potatoes, radishes, and turnips. Root crops need more potassium than the other groups in terms of good crop development.

GOOD TO KNOW

It's true that potatoes have the need for good potassium so they technically belong in the root crop category. However, if you're also growing tomatoes, you don't want to follow the fruit (tomato) group with potatoes when rotating crops. These two are in the nightshade family and they're a solid draw for what ails both plants.

Flower crops. The flowering crops are the legume family, which would typically place them into the fruit category. But legumes such as fava beans, green beans, and peas are special, so they deserve their own group. Legumes have the capability to fix atmospheric nitrogen and store it in their roots, so they add nutritional value to the soil.

Legumes are referred to as a "green manure" or "cover crop" because they can be planted specifically to add nitrogen to a garden bed. When you're done harvesting your legumes for the season, rotate the leaf crops into this bed since they could really use that nitrogen boost.

Fruit crops. This group includes plants that you harvest for their fruit, such as cucumbers, eggplant, melons, peppers, pumpkins, squash, tomatillos, and tomatoes. The fruit crops need extra nutrition, but phosphorus is the most important nutrient for the fruit crops. In fact, many a gardener has been frustrated over their tall, lush, but nearly fruitless tomato plants, only to figure out later that they've been fertilizing the bejeezus out of them—with nitrogen. Lots of nitrogen = lots of leaves.

What's the Point of Pruning?

What does pruning have to do with vertical gardening, you ask? Depending on what you've planted, maybe nothing and maybe everything. The answer also depends on your definition of pruning.

The dictionary defines pruning as:

1. To cut off or remove dead or living parts or branches of (a plant, for example) to improve shape or growth.
2. To remove or cut out as superfluous.
3. To reduce: as in prune a budget.

There are many reasons to prune in a vegetable garden, but major pruning techniques are also used on fruit trees and berries. Let's take a look at *why* we prune plants.

Stimulate new growth. Pruning a growing shoot stimulates new growth production. So if you're looking for some vigorous new growth on a shrub, prune it hard (a lot). Consider this type of pruning when you have a shrub that has a weak section of growth, such as at the back.

Restrict a plant's size. This can be especially important if you live in an area with restricted space, or you have a vertical or small-space garden. Gardeners living in urban and suburban areas almost always have to perform some pruning to keep trees and shrubs from outgrowing the yard, garden, or container.

Let in more light. If you have an extremely shady yard or you'd like to have more sun reaching the area under a tree for plants or lawn, careful pruning can let in a little extra sunshine.

Provide health and structural soundness. Any diseased, injured, dying, or dead branches should be removed for the health of the tree. Branches that rub together should be removed to eliminate potential damage to a main branch. Much of maintaining structural soundness in a tree is about careful pruning practices such as not "topping" trees. Topping can make the tree weak and susceptible to pests. It's also associated with slow death, even if it takes years for the tree to actually die.

Create special effects. Pruning for special effects is most often seen in formal-type gardens. They often take the shape of boxwood topiaries, or an apple tree that's been trained as an espalier (defined later in the chapter).

Encourage flowering and fruit. Pruning can coax growth spurs (produces the flowers and the fruit) to form on the branches. Strong flower buds are also encouraged to form due to pruning. Fruit trees can be lightly pruned in the summer, which will provide better air circulation around the fruit. This pruning results in less trouble with fruit diseases and faster fruit ripening.

Protect people, pets, and property. Trees that have been planted near homes, sheds, play structures, and other buildings pose a potential threat to human safety if heavy branches break off or the tree falls. They can also interfere with telephone or power lines. Proper pruning can keep people, pets, and property safe.

Keep evergreens proportionate. Pruning will keep boundary hedges under control. Evergreens benefit from light pruning as it keeps their foliage dense, and therefore, attractive.

Improve appearance. For many gardeners, pruning is about their plants' appearance. Removing little dead or unwanted branches, creating a pleasing shape, and removing suckers keeps plants looking neat and at their best. Plants that have grown out of balance with either the yard or their own growing pattern (such as stray and awkward branches) can be reshaped by pruning.

Most pruning is about working with a plant's natural growth pattern as it's developing, as well as maintaining mature fruit trees along with other tree and shrub species. One of the few exceptions is when you're pruning for special effects such as espalier (a pruning technique that trains fruit trees to grow flat along a fence or wall). That said, you'll be doing a certain amount of pruning in the vertical vegetable garden.

PRACTICAL PRUNING: TOMATO PLANTS

Should you prune your tomato plants or leave them to their own devices? There are some great reasons to prune them and one good reason to leave them alone. The answer has to do with the type of tomatoes that you have planted and your personal preference. Vining tomato plants that produce fruit all season long are called "indeterminate" tomatoes, and therein lies a terrific example of pruning in the vegetable garden. Determinate (bush type) tomato plants are rarely pruned because they reach a certain height and set fruit all at once.

Indeterminate tomatoes can grow to be 10' tall and become a veritable jungle. Some of the branches are actually "suckers" that just drain the plant's energy. Yes, they'll eventually flower and produce fruit, but maybe not before they outgrow the garden space that you've allotted them.

There are a couple of issues with all those branches and one of them is energy. Tomato plants have to use a lot of energy to produce those leaves and branches alongside the fruit. Pruning will alleviate some of that burden, and the energy will go to producing fruit.

An extra perk is that tomato plants stay slender and under control so they can be grown in a smaller area than a plant that's allowed to run rampant.

If there's a downside, it's that there will be less fruit produced because pruning involves removing would-be branches. This doesn't bother me with an indeterminate plant because they continue to give me fruit up until a hard frost.

If you're growing determinate plants, pruning may not be the right answer. Determinate tomatoes grow to a specific height and then produce their tomato harvest all at once. Most people don't want to limit the tomatoes grown on a determinate bush because you get one shot at fruit and then the show is over.

How to Prune Tomato Plants

1. Wait for your tomato plant to flower for the first time. The first thing you do is *start at the* ***bottom*** *of the plant* and find the first flower cluster. Using clippers or pruners, remove all of the lower branches below that first flower cluster. Don't panic, just do it. Soil-borne diseases love to attach themselves to that lowest branch and make their way up the plant. Plus, any branches below that first flowering one won't produce fruit anyway.
2. Now you're going to remove any suckers (extra forming branches) on the plant. These are side shoots that grow between the main stem and the branches. You'll recognize them as a pair of tiny leaves growing in the *V* (axil) of the stem and a branch. Of course they're only tiny in the beginning, and it's easiest to pinch them off while they're little.

   While they're tiny, just pinch them off with your fingers. If the suckers are a couple of inches long already, then use your pruners to remove them. You always have the option of leaving a few suckers to continue growing, and pinching off the rest if you'd like. Be sure to leave the terminal shoot (the growing top of the plant) alone.
3. The last tomato pruning practice comes at the end of the season. About 4 weeks before the first frost date in your area, go ahead and top the plant. Topping is to remove the plant's terminal shoot so that the plant will stop growing taller and begin sending all of its energy and nutrients to the last of the fruit so they reach maturity before the frost hits.

Pinching and Deadheading

When we think of pruning, it's the woody and vining plants that come to mind. We may picture piles of branches worthy of a gas-powered chipper-shredder. Pruning may not ever get quite that serious in your vertical vegetable garden; however, kiwifruits and grapes need pruning in order to train the vines.

Espaliered fruit trees will need to be trained along a trellis or a wall, as well. Grafted plants (like grapes) will need the shoots (called suckers) growing from below the graft union (swollen part of the plant) pruned off. You can gain control of the size and fruit production of those indeterminate tomato plants by pruning off the suckers there, too.

Even if your garden doesn't require the previously mentioned pruning techniques, there will be other versions of pruning going on: pinching and deadheading. Both of these practices are pruning techniques in their own right: you're removing part of the plant. But pinching and deadheading aren't the same thing. It's all about the timing.

When you remove flower buds and developing seed heads from cilantro, you're pinching. Pinching is done before the plants bloom to control the shape, size, or flower production. In the case of this herb, you want to stop the flowers from showing up because it sends a signal to the plant to stop growing—that its life is over. We try to put off this signal as long as we can because we want to harvest cilantro's leaves.

When you remove the spent flowers or seed pods from a plant, you're deadheading. It directs plant energy to new plant growth as opposed to the dying stuff. This is why deadheading flowering plants routinely gives you the maximum amount of blossoms.

What's Bugging You?

9

Yes, there's going to be trouble in paradise eventually. Did you know that there's a garden plan that can help limit this trouble to minor episodes as opposed to big disasters? This plan can keep your garden chemical-free and will still produce a bountiful harvest.

The plan is simple: grow a healthy garden. This means following the practices I discussed in earlier chapters, such as starting with good soil, choosing the right location, and taking advantage of disease-resistant plants. The bad news is that no matter how hard you try, at some point some pest is going to come along and give you and your garden a little grief. The good news is that Mother Nature anticipated this and she's got your back.

There's an army of beneficial insects that come free just by planting your garden and using the organic practices that I'll discuss in this chapter.

Organic Pest Control

The term *organic* means different things to different people. It sounds pure, healthy, and sustainable—as it should be. But even practices, products, and things that are organic in nature can be dangerous for *something,* even if just for the pests they smother. For the sake of argument, when I use the term *organic,* I'm speaking in terms of something that's the least capable of harming the environment, humans, and animals.

I probably don't need to explain *why* organic gardening is the best way to go for your vertical garden, wildlife, your family, and the environment. I do get asked if I actually have success with organic gardening practices. The answer is that I honestly, seriously, most definitely get excellent results without using pesticides and herbicides (chemical anything) in my gardens. I completely understand that this may not be possible for everyone. Although I can't help but wonder if many people who don't believe that it works have actually tried it for themselves or tried it for any length of time. Because if you've been plying your yard and garden with synthetic chemicals—including fertilizers—for years, there's going to be a period of adjustment.

GOOD TO KNOW

The key to successful organic gardening is that you have to be willing to experiment with different techniques to find the one(s) that works best for you.

Every plant and creature will have to readapt to the new lifestyle. The beneficial-insect-versus-pest-insect balance will certainly be out of whack for a bit, but the good guys always end up winning (but you knew that, right?).

Let's talk about what you can do to keep pests at bay organically, and then we'll move on to the least-toxic controls. If you have to take it a step further, we'll get into how to do it safely.

Beneficial Insects

As far as we gardeners are concerned, with insects there are the good guys and the bad guys. It might not be very politically correct of us, but there it is. Like knights to the queen, some insects are born plant protectors and some see our gardens as a veritable smorgasbord.

Insects such as Colorado potato beetles, snails, slugs, aphids, cabbage worms, and coddling moths are just a few on the "Insects *Least* Wanted" list, while beneficial insects that are the pollinators and predators are certainly on the "Insects *Most* Wanted" list.

Predatory beneficial insects won't eradicate every bad insect in the garden, but they take care of more than enough to balance out the equation quite sufficiently to establish beautiful, healthy plants and only a few munched-on leaves. With these natural soldiers on your side, you can eliminate (or drastically reduce) the use of potentially harmful synthetic pesticides.

Beneficial insects fall into two categories: pollinating and predatory. The pollinators make it possible for the garden to produce a vegetable bounty. The predators carry the heavy artillery and diligently devour the bugs that devour your garden. Your best bet is to invite as many beneficial insects from both classes into your vertical garden as possible.

Who's Who?

Several years ago my husband and I were at a local nursery and we overheard a woman asking where the pesticides were to get rid of an insect that she was finding all over her garden. She had brought one with her in a plastic baggie to show the nursery attendants in order to find the right poison to rid her garden of the offending critters.

From where I was standing, I couldn't quite make it out, but the little guy looked very familiar. Much to my husband's embarrassment, I stalked the woman as she perused the aisles looking for just the right weapon. When I finally got close enough to see the orange insect paddling his legs against the slick plastic, I commented (my husband says I shouted, but I don't think so), "Hey, that's a soldier beetle—don't kill him! He's the cavalry!"

The woman said that she found many of them among her roses and I asked her if she had also found a lot of aphids on her roses as well. She assured me that aphids were covering the plants—bingo! The soldier beetles had heard her roses' cry for help and showed up just in the nick of time.

The obvious moral of the story is to learn to recognize the difference between friend and foe. Identify your local helpful critters so that you don't accidentally wipe out your own troops. You'll find beneficial insect images online, in organic gardening books, and in field guides.

GOOD TO KNOW

An excellent one-card guide called *Mac's Field Guide Bug Identification Page* is available at Amazon.com. It's a laminated sheet with pictures of beneficial insects on one side and garden pests on the other. If you're really not sure what bug you're looking at, feel free to catch one in a jar and bring it down to a local nursery, or better yet, your Cooperative Extension office for proper identification.

Pollinating Insects

Following is a basic list of the insects that pollinate your vegetables and fruit. I should point out that some predatory insects are useful pollinators, too.

Blue mason bee (Orchard bee). These little, docile blue bees are often mistaken for flies. Many a gardener has wondered why flies are enjoying their roses, never suspecting that they're actually looking at an effective, early spring pollinator. Gentle, Blue mason bees don't make honey and they don't have an aggressive bone in their bodies. In fact, the males don't have a stinger at all and although the females have them, without any honey to protect, they aren't inclined to use them. Blue mason bees are just one of our many pollinating native (non-European) bees.

Butterfly. Butterflies may not be as effective a pollinator as some because the pollen just doesn't stick to their bodies well. That said, I'd invite them into my garden regardless.

Fly. Flies? Really? Okay, to be fair we aren't talking about houseflies here, but rather black soldier flies, tachnid flies, syrphid flies, and bee flies. It might surprise you to learn that flies are second only to bees as pollinators!

Honey bee. While there are many different bee species, the European honey bee reigns supreme in the pollination department. Currently, we're experiencing a worldwide honey bee loss due to what is called "Colony Collapse Disorder." Experts everywhere are searching for the answer to what is threatening the very existence of our best pollinating insect. This is why organic gardeners are doing their part by steering clear of pesticides in the home garden.

Lacewing. As adults, lacewings are dressed in lovely, bright green with gossamer, fairy wings. Adults do the pollinating and their children eat all of the things that go crunch in the night. They're truly one of the most effective beneficials that you can have visiting your garden.

Moth. Moths are the pollinators of the night-blooming plants. Although there is a daytime moth that's probably fooled you more than once: the beautiful hawk moth or hummingbird moth is typically mistaken for a hummingbird. Even if you don't remember seeing one up-close-and-personal, you've surely met them as young ones—the tomato hornworm, profiled a bit later in the chapter.

GOOD TO KNOW

All tomato gardeners dread finding hornworms on their plants, and rightly so. Hornworms seem to consume plants in record time. However, they do deserve a little respect because as adults, they're important pollinators for certain plant species. This leaves us wondering what we should do when we find them among our tomato plants. My solution is to sacrifice two tomato plants and toss the rest of the hornworms to my chickens (gruesome, I know). That's right; being a pollinator advocate I actually have a couple of tomato plants that are situated away from my tomato bed and if hornworms invade those, I let them be.

Wasp. Many different wasp species are very effective as pollinators—some are even better at it than bees. In fact, some are responsible for the pollination of several plants, including fig trees.

Natural Predatory Insects

Depending on the species, many of these predatory insects double as pollinators in the garden. This list isn't exhaustive by any means. For more information on local beneficial insects, contact your local Cooperative Extension office.

Assassin bug. These predators prove that you don't have to be good looking to get the job done right. What they lack in attractiveness they more than make up for in appetite and speed. Favorite meals include Colorado potato beetles, cabbage worms, aphids, tomato hornworms, cucumber beetles, cutworms, Japanese beetles, and caterpillars.

Dragonfly and damselfly. Adult dragonflies and damselflies are tenacious predators. As carnivorous grown-ups, they're both superior hunters-of-the-skies and snatch their prey in mid-air. They're also one of the fastest insects in the world. As nymphs living in the water, they have insatiable appetites for water insects such as mosquito larva.

Green lacewing. Also called the aphid lion, these predators use a pair of curved mandibles (jaws) to harpoon aphids and suck the life out of them. They also eat other soft-bodied insects such as mites, mealy bugs, spider mites, whiteflies, scale, and thrips. The adult lacewing is a pollinator. Their offspring are tenacious, too; green lacewing larva can eat 60 aphids per hour.

Ground beetle. Pretty shells, large mandibles, and a voracious appetite pretty much describe ground beetles. With a tendency to hide under plant debris on the ground, you may not notice them, but at night you can be sure they're on the hunt.

Hoverfly. Also called syrphid flies, the larvae feed on soft-bodied pest insects such as mealy bugs, aphids, maggots, and caterpillars. They resemble a little bee and have one pair of wings, yellow-striped bodies, and huge compound eyes.

Ladybug. These assassins cloaked in a red Volkswagen's clothing will each consume 5,000 aphids by the time they die. Other ladybug prey includes bean thrips, mites, chinch bugs, Colorado potato beetles, and asparagus beetles.

Ladybug larvae. These little dudes are black and orange-red with a prehistoric alligator look. These spiny little creatures aren't much to look at, but they can eat as many as 50 to 70 aphids a day.

Leather-winged beetle or soldier beetle. These slender guys are long and orange with dark wings. If your roses are covered in aphids, the super-hero soldier beetles aren't far behind! Give them a chance to come to the rescue.

Minute pirate bug. They may be tiny (like ⅛") but these predators help control small caterpillars, aphids, mites, and thrips. They move lightning-fast when they spot a good meal, piercing it with their needle beaks.

Praying mantis. Although mantids (plural for mantis) are big consumers that don't always discriminate between good and evil, they certainly eat garden pests. On the other hand, they sometimes grab a good guy or two in the process.

Spider. People are often repelled by these eight-legged creatures, but spiders are friends to the garden. After an insect is caught in a spider's web it is quickly wrapped up by his host and injected with a venom that liquefies the insect and the spider just sucks him down. They eat more insects in the garden than birds and they help out with pest control year-round.

Spined soldier bug. Another master of the harpoon attack, spined soldier bugs inject a paralyzing substance into their prey and feast on the juices. Potato beetles, tomato hornworms, caterpillars, saw-fly larva, and cabbage worms end up as this predator's dinner.

Trichogramma wasp. This wasp doesn't bite … humans, that is. It's just one in a group of parasitic wasps that lays its eggs inside the larvae of garden pests such as cabbage worms, cutworms, and borers. Once inside, they dine on the internal organs of their host. The aphid ends up mummified as baby wasps spin a cocoon in there, pupate, and finally emerge as adult wasps.

PLANTS POLLINATORS CAN'T RESIST

How do you encourage beneficial insects to hang out at your place? You whet their appetite by planting any of the following irresistible plants around your vertical vegetable garden. They don't have to be growing in the same beds—just adding them to your yard will be enough to send out the perfect invitation.

Alyssum
Bee balm
Black-eyed Susan
Blazing star
Candytuft
Catnip
Coneflower
Coreopsis
Cosmos
Dill
Fennel
Floss flower
Goldenrod
Lavender
Lovage
Marigold
Mexican sunflower
Monarda
Queen Anne's lace
Red clover
Scabiosa
Stonecrop
Summer savory
Sweet Cicely
Tansy
Thyme
Yarrow
Zinnia

Common Vegetable Pests

Now let's take a look at some insects that do damage to your vegetables, and the predators that control them.

Aphid. These teeny, pear-shaped, plant-sucking ladies can show up in colors of green, brown, black, gray, red, and yellow. In small numbers, their damage is minimal, but infested plants can suffer new growth to be curled and distorted, which can stunt them. Natural aphid predators are praying mantids, lacewings, minute pirate bugs, hover flies, damsel bugs, big-eyed bugs, ladybugs, assassin bugs, and spiders. Just about everyone enjoys them for lunch!

GOOD TO KNOW

If it seems that a bazillion aphids are born overnight, it's because they are—literally. Aphids are all females, and every aphid is born pregnant. However, aphids give birth once they are mature adults, which is about 10 days after they're born. During the warm spring temperatures, special aphids called "stem mothers" emerge from wintering-over eggs. The stem mothers give birth to live daughters, who are also pregnant; no male is necessary. It's truly a superior race of females gone terribly wrong. The phenomenon continues with every generation of female aphid until the end of the season, when things get really interesting. The aphids begin to produce daughters and sons. These sons mate with the current generation of female aphids and those females lay eggs on bud scales to winter-over … and the cycle continues.

Colorado potato beetle. Dressed in dapper, unassuming, pin-striped suits and about half the size of your thumbnail, these rather cute fellows *look* relatively harmless. But make no mistake, they have a voracious appetite and will defoliate plants. Ladybugs, damsel bugs, parasitic wasps, assassin bugs, tachinid flies, spiders, lacewings, and praying mantids are all natural controls for the Colorado potato beetle.

Cucumber beetle. There are a few different species of cucumber beetle. The yellow-green ones have black spots on their wing covers, and the yellow ones have three black stripes running down their covers. In any case, the damage is the same: eating flowers off of crops, making holes in leaves, and consuming germinating seeds. Natural predators for the cucumber beetle are tachinid flies, assassin bugs, and parasitic wasps.

Earwig. Otherwise known as the infamous "pincher bug," these pests are hard to mistake as they have large, ominous pinchers on their back end. Earwigs chew on seedlings, leaves, flowers, and maturing fruit. It may surprise you to know that they play a more positive role in the garden as well. They also enjoy decaying matter (helpful for creating compost) and act as a predator by eating aphids. Praying mantids, assassin bugs, and tachinid flies are all enemies of the earwig.

Japanese beetle. Japanese beetles are a metallic, iridescent bronze/green color and have plump bodies. They eat the tender part of the leaves, leaving only a "skeleton" behind. These beetles also have no problem eating the flowers while their grubs (young) eat away at the roots of your lawn. Parasitic wasps are the primary predator for Japanese beetles.

Mexican bean beetle. You may dismiss these beetles in the garden because they look an awful lot like ladybugs, only they're bigger and on your bean plants. Orange with 16 black spots on their wing covers, like Japanese beetles they feed on leaves, leaving only a skeleton behind. Stems of plants and the beans are often consumed, as well. Praying mantids, ladybugs, assassin bugs, tachinid flies, parasitic wasps, and minute pirate bugs will be happy to take care of them for you.

Snail and slug. It's hard to miss these slimy pests that creep along the ground leaving a slimy residue behind. Whether they carry a shell on their backs or not, they're a garden demolition crew. Seedlings and new plant growth are the usual victims of snails and slugs. As opposed to insects, frogs, toads, snakes, moles, and birds are the best at eradicating these pests. In fact, these critters will help control all garden pests.

Squash bug. Squash bugs are brown or gray with long antennas and a flat back that resembles a shield. They feast on plant juices by piercing them and sucking away. You may notice small dots on plant leaves, only to have the leaves turn yellow, wilt, and die a short time later. Natural predators are spiders, praying mantids, and tachinid flies.

Squash vine borer. These 1½"-long moths have black and red bodies with clear wings. It's their white young that bore into plant vines (usually in the cucurbit family) to eat the flesh. Parasitic wasps and ground beetles can help.

Tomato hornworm. These guys are incredibly chameleon-like in their ability to fool the eye. But once you come across one, you'll never forget it. Fat-bodied and the same green shade as your tomato plants, these 3"–5"-long caterpillars defoliate plants in just a day or two. Plants that are hosts to hornworms have entire leaves missing (sometimes the stem is left behind). You may notice ¼"-long, dark green "pellets," which is (you guessed it) caterpillar poop. Parasitic wasps and assassin bugs will take down this crop destroyer. (See the sidebar earlier in the chapter for the good side of the tomato hornworm—as a valuable pollinator for some plant species.)

Helpful Wildlife

Aside from helpful insects that seek and destroy what's bugging you in the garden, there are others in the animal kingdom that are perfectly willing to lend a hand, as well. Bluebirds, chickadees, swallows, purple martins, wrens, and nuthatches are insect-eaters that love the taste of caterpillars, stinkbugs, grasshoppers, aphids, and cucumber beetles. Entice them into taking up residence in your yard with bird feeders, nesting boxes, running water, and birdbaths.

The summer song of frogs and toads isn't the only reason to have them hang out at your place. Both critters are top-notch for natural insect control. Frogs are fabulous, but most need something that resembles a pond in order to keep them around. That said, little tree frogs that live in the damp spots in our garden seem to do just fine. Toads, however, are quite happy with water-filled plant saucers (that have been sealed) buried halfway into the soil.

Least-Toxic Pest Controls

Products that have proven to be effective, yet cause the least environmental disruption are referred to as "least-toxic." They're popular among home gardeners because they break down quickly after being applied and are harmless after a few days. They're non- or low-toxic to pets and people, too.

Some prepared formulas that you'll find on the market are also supposed to be safe for beneficial insects. That's a harder claim to prove, plus the other school of thought is that anything that kills something can't be considered entirely nontoxic. That said, if you need something a little stronger, these products are the better choice. All of the following products can be found in garden centers across the country, as well as online.

DOWNER

Before you spray, dust, or apply your vertical vegetable, herb, berry, or fruit garden with *anything,* make sure the product is safe to use on edible plants.

Diatomaceous Earth

This product is inexpensive, effective, and nontoxic to people, pets, and wildlife. Diatomaceous earth (DE) destroys ants, earwigs, slugs, beetles, ticks, fleas, cockroaches, and bedbugs. All you do is spread it around the ground wherever you've seen these unwanted pests. As they move across the powder, it sticks to their feet and legs and works its way into their joints and exoskeleton.

This is bad news for these insects because DE is made up of crushed fossilized skeletons of diatoms and algae. It's jagged and sharp, so it works like little pieces of broken glass and scratches up the insects' bodies and then dries up their fluids.

I've heard a slight variation to this theme which says that instead of scratching up their bodies, the DE is absorbed into the bug's breathing tubes as well as their joints, eyes, and so on. In any case, the bugs quit eating and die soon after.

DE feels like a powder and is completely safe for humans to touch. That said, you don't want to breathe it into your lungs, so wear a dust mask when you apply it. Also, there are various grades of DE; there's one for swimming pool filters and one that's horticultural grade. You want the horticultural one. I should mention that with DE you don't get an instant kill; it may take days or weeks to be rid of the above pests completely, so a little patience is necessary.

Iron Phosphate

To eradicate slugs and snails, iron phosphate is my method of choice. You'll find it under the brand names Sluggo and Escar-Go! It's safe for pets, people, and wildlife, and is very attractive to the gastropods. Thankfully, this snack proves to be their last, as 3 to 6 days after ingesting the Sluggo, these critters crawl off and die.

During this time you may see them still creeping along the soil, but they can't actually eat, which is the beauty of iron phosphate. What the critters don't eat breaks down and become fertilizer for your plants; it's a win-win!

GOOD TO KNOW

One of the most effective, completely nontoxic pest controls is your own hands! Hand-picking pests and disposing of them goes a long way toward a healthy garden.

Insecticidal Soaps

Premixed insecticidal soaps (considered nontoxic) can be found at nurseries, garden centers, and online. Spray soaps aren't a preventative measure; they control pests after they've already arrived on your plants. Soaps coat the critter, which causes the insect's cells to collapse. In other words, they kill the bugs that are bugging you. Also once the soap has dried, it's no longer effective.

Sometimes you just can't beat a lesson from a little old-fashioned soap. Soft-bodied little suckers (literally) such as whiteflies, aphids, mites, and leaf miners are easily done in by a simple soap spray. You can try this in your own garden by mixing 3 to 5 tablespoons of liquid dish soap and a little vegetable oil to a gallon of water. Pour the mixture into a hand-held spray bottle or garden sprayer and use it on infested plants during the morning hours when the sun is low and temperatures are cool. Make sure to spray under the leaves as well.

It's always smart to try a test spot with a plant leaf before spraying an entire plant with anything. Most won't be adversely affected, but you never know. If the leaf you tested acquires brown spots or the edges turn brown, look for a different solution.

Bacillus Thuringiensis (Bt)

Bt is a naturally occurring bacterium that can be extremely effective in controlling plant-devouring caterpillars. This is another control that isn't a "knock-down" solution; it works slowly, but surely. It's a very specific bacterium that targets certain caterpillars. After they snack on it, the Bt causes them to stop eating, and they end up dead. Bt is safe for crops, pets, people, and beneficial insects, although it may kill nontargeted butterfly larvae, as well.

Horticultural Oils

Horticultural oils coat-to-kill insects just like the soap sprays—but they work longer. They're environmentally friendly and safe to use around mammals. Horticultural oils are useful for controlling aphids, mites, scales, leaf rollers, moths, and caterpillars. These oils effectively kill eggs along with the adults.

Dormant oils are applied in late winter on trees and plants before they leaf out, while summer oils are applied when temperatures outside are over 40 degrees but under 85 degrees. Before you apply oil (petroleum) products, be sure to choose the right horticultural oil to the right pest.

One last thing: do your plant research before using oils for pest control. Some plants such as the Japanese maple are known to be extremely sensitive to these products. The summer oil sprays are much lighter, so this may be the better choice for sensitive plants.

Pyrethrum

Pyrethrum is a natural insecticide derived from chrysanthemum flowers. It attacks the central nervous system of aphids, mosquitoes, flea beetles, moth larva, thrips, small caterpillars, and more. It's important to mention that pyrethrum doesn't know the difference between the bad guys and the good guys, however. It's safe for people, but moderately toxic to certain mammals. It's not safe for water and fish, so be aware of runoff or if you're close to a body of water (creek, pond, etc).

DOWNER

Every precaution should be taken to keep *all* pesticides and herbicides from reaching waterways. These chemicals can be incredibly damaging for both aquatic animals and other wildlife.

Breaking Out the Big Guns Safely

There may come a time when you feel that you've tried every organic technique you can think of and the pest (or weeds) still persist. I hope you don't get to that point (as I'm sure you don't either), but it happens. It's said that home gardeners often use more pesticides per square foot than commercial farmers. If this is true, then it's probably because many people are of the mind-set that "if a little works well, then more is better." This is not only untrue, but it can lead to some serious chemical misuse and home gardeners are responsible for the misuse of pesticides on their property. You may be thinking, responsible for *what* exactly?

Chemical pesticide and herbicide misuse can result in injury to animals, the environment, and desirable plants (including your neighbor's), and can leave you over-exposed to the product, as well. So before you grab a pesticide (or herbicide) and start blasting the place out of frustration, let me share with you some safe and sane guidelines:

- First, identify the pest you intend to control so that you can choose the appropriate pesticide. If you find that you have a product choice, choose the one that is selective about which pests it controls, has the quickest breakdown, and exhibits the lowest possible toxicity to people and animals.
- Keep children and pets out of the garden and immediate area where you're applying pesticides or herbicides, and don't allow them into the area until the product is completely settled or dried.
- Be very certain that the product you've chosen is safe to use around or on edible plants before applying it to the vegetable, herb, or berry garden.
- Always wear a face mask, long-sleeved shirt, gloves, and safety goggles whenever you apply chemical pesticides. The idea is to avoid inhaling the products or letting them come into contact with your skin, face, or hands. After handling the product, don't touch your face until you've washed your hands. You also shouldn't be eating or drinking at the time of application.
- If you're applying an herbicide, if possible, consider the idea of "painting" it onto the unwanted plant instead of spraying. Keeping the product limited to only the affected plant not only lessens the chances of it coming into contact with the unaffected plants surrounding it, but also reduces the risk of spreading it into the environment.
- Always read the label on a product before you use it. Product labels contain important directions, legal use, and safety information on handling, applying, and storing these products correctly. Any brochures or flyers that accompany the box are just as important for addressing additional instructions or limitations.

- Apply these products on a day that's wind-free with a moderate temperature.
- Thoroughly clean all equipment such as hoses, containers, tanks, and nozzles after using the product. Wash any clothing you wore during the application separate from other laundry. Your hands and face should also be washed immediately afterward.
- Keep pesticides from reaching swimming pools or other bodies of water such as streams or ponds. Don't spray near birdfeeders and pet bowls.
- Store all chemicals in their original packaging in a locked, dark, cool place. Improper chemical storage is the reason that children are the major group of nonagricultural victims of pesticide poisoning.

To be clear, *any* pesticide is capable of harming the good guys (beneficial predators) just as well as the bad guys. If you share many gardeners' concerns to protect beneficial insects, use pesticides only when you can find no other practical and effective pest control.

Understanding Pesticide Toxicity

Chemicals are assigned a toxicity category as part of their registration. The symbol that expresses that assignment to the public is represented by a "signal word," which is *Caution, Warning, Danger,* or *Danger Poison.* Here's what these signal words mean to you:

- **Caution:** This is a product of the lowest toxicity. A product labeled with this word can range from relatively nontoxic to slightly toxic. The approximate human lethal dosage is an ounce (in the case of a slightly hazardous product) or more (say over a pint for a fairly nontoxic product).
- **Warning:** A moderately toxic or hazardous product will be labeled with a *Warning* label. The human lethal dose is approximately 1 teaspoon to 1 ounce.
- **Danger:** Products labeled *Danger* are highly hazardous and are best kept out of your home entirely. The lethal human dose is pesticide-specific, so read the label very carefully.
- **Danger Poison:** Any chemical labeled *Danger Poison* will be accompanied by a skull and crossbones image. It alerts you to the fact that the product is *highly toxic.* The lethal human dose is anywhere from a small taste to 1 teaspoon. Your best bet is to steer clear of this type of product.

Integrated Pest Management

Integrated Pest Management (IPM) evolved from the University of California entomologists practicing "integrated control" in the 1950s. In 1972 it was stated as national policy in the United States.

The entire premise of IPM is based on beginning with the least toxic pest remedy and working up in graduated steps, technique by technique, and stopping when something works. It's a good technique for organic gardeners and those who are living a sustainable lifestyle. IPM relies on the gardener or farmer to gather information and monitor the garden. IPM doesn't prohibit pesticides, but seeks to solve pest problems before reaching the pesticide level of control. IPM users often encourage "spot treating" (applying the pesticide to only the plant that's affected instead of wide-spraying) when it comes to pesticide use in the garden.

Following is an overview of how IPM works.

Determine whether you really have a pest problem. The first thing gardeners have to realize is that the mere presence of an insect does not a problem make. What kind of bug is it? Is it damaging your plants? How do you know? Look for evidence of the offender's natural enemies. If you have aphids, look for soldier beetles or ladybug larva. Where there are bad guys, there are bound to be good guys. It may turn out that you don't need to do anything at all. Learn to recognize your local insects—the good *and* the bad. If you can't identify it, bring one in a jar down to your local Cooperative Extension office, or at least give them a call.

GOOD TO KNOW

It's amazing how effective a good water blast is, aimed at a group of aphids. It knocks them down, breaks vital parts, and they're never the same after that. Which means that they can't ever climb up your plants again. (Is it wrong of me to smile as I write this?)

Assess the real damage. Are there just a few cosmetic blemishes on a few fruits? Or has all your baby spinach become a snail family's dinner? Has the offender moved its entire family, cousins, and friends to your place? Or do you have just a few vacationing? Think about it, because there can be a vast difference between the two. We're programmed to be intolerant of any pest that crosses our path. But if you have only slightly blemished fruit, the natural balance is probably pretty stable. In fact, if that's all the worries you have, this is what we call successful gardening.

Access your tolerance level. Everyone has his or her own tolerance level. For instance, you may decide that the damage isn't a very big deal, yet someone else may feel that the particular plant isn't worth the hassle. In that case, another choice may be to simply rip out that particular plant and replace it with a more pest-resistant species.

The IPM control program uses part or all of the following pest controls, which are cultural, physical, biological, and least-toxic. Although gardeners will utilize the *first three* as graduated steps, they're often used in tandem:

- **Cultural controls:** These practices include using plant varieties that are disease-resistant, instituting crop rotation, tilling and cultivating soil, changing the planting or harvesting time, following proper watering and fertilizing guidelines, and utilizing companion planting as well as good sanitary practices.
- **Physical controls:** Good examples of physical pest controls include the use of diatomaceous earth, copper strips, sticky traps, and hand picking.
- **Biological controls:** The use of beneficial insects, competitive insect species, pathogens, predators, and parasites are all biological controls.
- **Least-toxic chemical controls:** These are products that are the least disruptive to the environment, people, and animals. They include insecticidal soaps, oils derived from plants (neem tree and mint), pyrethrum, Bt, boric acid, borax, and borates.

Vegetables and Fruit That Enjoy Growing Up

4

The simple act of planting and tending a garden is both a joyful and rewarding experience. Harvesting food for you and your family is the icing on the cake!

Part 4 is dedicated to planting, tending, and harvesting the vegetables, fruit, and herbs that thrive in a vertical setting. I also discuss those plants that may not physically grow up (vines), but are specially suited for vertical containers. I include a section for each plant profile that discusses planting, tending, harvesting, and the best-bet varieties.

(Literally) Vertical Vegetables 10

In this chapter, you find those vegetables that grow up naturally for planting on trellises, frames, and other supportive structures. I've got planting directions, tips, and great varieties for your vertical garden.

The plant variety lists are a good place to start; however, I encourage you to try other varieties that you find interesting. There's more than one way to successful gardening.

Beans (*Fabaceae*)

You've got a lot of choices when if comes to growing beans vertically. Most bean species are warm-weather crops, with the exception of fava beans, which favor the cooler weather. In any case there are a few differences with each, so I've broken them into their respective categories. One commonality is that they're all legumes and will take nitrogen from the atmosphere and fix it into their roots—a welcome perk for the vegetable garden.

Pole and bush beans (*Phaseolus vulgaris*). This group is the most widely planted in home gardens. Depending on the variety, both pole and bush beans can be grown as snap, shelly, or dry beans. Bean pods' colors on the vine vary and may be bright green, yellow, purple, and purple/green striped (although those varieties with purple pods do turn green while they're cooking).

The bean group belongs in both the vertically favored veggies (featured in this chapter) as well as the vertically challenged veggies (featured in Chapter 11), because there are as many bush bean varieties as there are climbing varieties. The directions in this chapter apply to both types of beans, so I won't repeat them in the next chapter. Varieties mentioned here will be mostly pole varieties, but we'll have some bush beans for containers, too.

GOOD TO KNOW

Pole beans are the obvious choice for vertical gardening, but the bush varieties can be grown at the feet of other crops. I prefer the pole types because they not only produce longer, but I find that they have more flavor than the bush varieties. On the other hand, bush beans mature faster for an earlier harvest.

Edamame or soy beans (*Glycine max*). Asia's gift to America, the soybean's short, fuzzy pods grow best in a humid heat zone as opposed to a dry one. When green soybeans are shelled and served, they're referred to as edamame.

Fava beans or broad beans (*Vicia faba*). This is the exception to the otherwise heat-loving beans. Fava is a cool-weather crop that's often grown as a green manure and cover crop. Most people harvest the immature pods for the kitchen.

Scarlet runner beans (*Phaseolus coccineus*). These beans grow very much like average pole beans, but they produce showy flowers as compared to a typical pole bean. Scarlet runner beans are fast growers and are often used as beautiful, living screens for the season.

Planting Beans

Feel free to start your beans directly in their outdoor bed in the early spring, when the soil reaches a constant 60°F. If you plant them earlier, not only will they just refuse to show their seedling heads, but the beans will end up rotting in the cool, damp soil. As far as starting them indoors, I typically don't, but have on occasion. Beans aren't thrilled with having their roots disturbed, but I've started them several weeks early indoors and they've transplanted well.

Find a spot in full sun and plant them about 1" deep and 3" to 4" apart in average soil. Beans aren't demanding and will grow in soils with little nutrition as long as it's not a bog; they like their feet in a well-drained site. Succession plant bush beans by planting another group (and then another) 7 to 10 days apart.

Tending Beans

Legumes eventually add nutritional value to the soil because of their nitrogen-fixing nature. But they don't do this while they're very young. You can offer them a balanced organic fertilizer or some fish emulsion every few weeks of the growing season. But when I give them an extra boost, I do so with a light hand, so as not to encourage too much lush, green growth.

Harvesting Beans

Your bean harvest will depend upon the bean type. The first place you'll consult on harvesting is your seed packet, because the days to harvest will let you know when you're getting close. In fact, the day that you plant them, go mark the day they "should" be ready on your calendar so that you'll have a head's up when you're getting toward the finish line. Try to harvest mature beans as quickly as possible in order to keep the fruit production coming.

Your green beans and scarlet runners will be ready anywhere from 50 to 80 days (check your packet for specific varieties). Soybeans will mature anywhere between 60 to 95 days. Fava beans need the longest growing time at 120 to 150 days.

If you're growing a dry bean variety, you'll want to wait until the pods are completely dry on the plant, to the point of shattering open themselves, before harvesting (84–100 days). If you're after shelly beans, harvest the pods several weeks before they're fully mature.

Best Bets: Bean Varieties

Pole beans are an obvious choice for vertical vegetable gardening, but I also love them because they continue to climb and produce all season long (unlike the bush varieties, in which the pods ripen all at once). That said, if you're looking forward to a bountiful bean crop, be sure to plant plenty of them so that you have many pods maturing at the same time.

Blue Lake Pole. Snap bean; smooth, dark green pods; mild, sweet flavor; 63 to 75 days to harvest

Cherokee Trail of Tears. Snap or dried bean; green pods with purple overlay and black beans; meaty flavor; 85 days to harvest

Dragon Tongue. Bush bean; extremely versatile as a snap, shelling, or dried bean; butter-yellow with bright purple stripes; juicy and tender; 60 days to harvest

Envy Soya (Envy Soybean). Soybean (edamame); shelly or dried bean; sweet, nutty flavor; 80 days to harvest

Golden Child Filet. Bush beans; 4", lovely flavor and texture; 55 days to harvest

Jacob's Cattle. Dried and shelly bean; white and red speckled; full-flavored; 95 to 100 days to harvest

Kentucky Wonder. Snap bean; huge harvest; smooth, silvery-green pods; old-fashioned beany flavor; 67 days to harvest

Lazy Housewife. Snap bean; big producer; stringless and flavorful; 75 to 80 days to harvest

Provider. Snap bean; popular and early because it germinates in cool soils; round, straight pods; stringless, with good flavor; 51 days to harvest

Rattlesnake Pole. Snap and dried bean; dark green pods with purple streaks; excellent flavor; 73 to 90 days to harvest

Romano Italian. Flat pods; great flavor; harvest when they reach 4"; 70 days to harvest

Royal Burgundy. Bush bean; 5"; dark purple pods and beige seeds; delicious beany flavor; flowers keep coming for a long harvest; 60 day to harvest

Sweet Lorane. Fava; tasty as a shelly and excellent as dried; bean seeds are light tan and have a chickpea flavor; 100 days to harvest

Taylor Dwarf Horticulture. Shelly or snap beans; pinkish-tan beans with red splotches; good pod flavor when harvested young but mostly used as a shelly bean; 68 days to harvest

Top Notch Golden Wax. Snap bean; golden, tender yellow pods; excellent flavor; 50 days to harvest

Cucumbers (*Cucumis sativus*)

Cucumbers, affectionately known by gardeners everywhere as simply "cukes," come in many forms: bush variety, vines, lemon-shaped, yard-long types, and burpless (low or no cucurbitacin, which adds bitterness and makes people burp after ingesting). If you've never grown cukes before, you may be surprised at the amazing number of varieties as you thumb through your spring seed catalogs.

Varieties are broken down into specialty categories, which are generally slicing, cooking, and pickling. Some of these varieties are dual-purpose and will cross over into several categories. Who knew?

Planting Cucumbers

Cucumbers can either be started early indoors or directly planted into their permanent bed in the spring. Indoors, start them 3 weeks before the last frost date in your area. Plant them ½" to 1" deep in a soil-less seed-starting medium and set them in a warm spot.

Cucumbers don't enjoy the transplanting process, so they should be handled carefully when planting them into the garden. They're good candidates for being started in little biodegradable pots that can be planted directly into the ground along with the seedlings. Cuke seeds can be planted directly into the garden bed once outside temperatures reach about 70°F. Plant seedlings 18" apart if you're training them up a vertical structure.

Tending Cucumbers

You'll get the largest yield from plants that are grown in fertile clay soils that have been amended with humus or other organic materials. Like all vegetables, they're happiest with an occasional application of compost or composted manure.

Cukes enjoy a moderate amount of watering up until they begin flowering. When the blossoms appear, they appreciate more water until their fruit is harvested. Ideally, they should be watered at soil level as opposed to overhead, as they're susceptible to mildew.

GOOD TO KNOW

Many gardening conversations revolve around slicing cucumbers and why some gardeners enjoy fabulous flavor while others end up with bitter fruit. Some gardeners blame it on hand-pollination (a technique where the gardener takes some pollen from a male flower and brushes it onto the female flower by hand). Let's set the record straight. Bitter cucumbers are all about cultural influences, not pollination—although some cukes that are grown specifically for pickling may be bitter if eaten raw. If plants are allowed to dry out instead of being kept evenly moist while the fruit is in production, or if soils fluctuate between very dry and wet, you run the risk of bitter fruit.

Harvesting Cucumbers

Cucumbers mature quickly, so check the vines daily for ripe fruit because if it's left on the vine and completely matures, the whole plant is signaled to halt production. How do you know if your cukes have reached that point? When the blossom end turns yellow, the fruit is overripe. They should be harvested when they're bright green—and don't forget that the smaller fruits are the most flavorful.

Best Bets: Cucumber Varieties

Cucumbers seem to make their way into nearly every food-related event. Share your bounty with pride by serving unique, home-grown cucumbers at your next barbeque or luncheon. My favorite "show off" variety are the Lemon cucumbers.

County Fair. Early producer with high yields; extremely disease resistant; great for slicing and pickling; they mature at 6" to 8" long, but harvest them around 3" to 4"; 51 days to harvest

Diva. Prolific plant; excellent flavor; harvest when fruit is 4" to 5"; 58 days to harvest

Homemade Pickles. Vigorous plants; solid and crisp fruits make big, juicy pickles; harvest them at 2" to 5" long; 55 days to harvest

Lemon. Small, round, yellow, and crunchy; good cuke flavor and little nuttiness (no lemon flavor); 3" to 4" round (lemon size); great for slicing and pickling; 67 days to harvest

Marketmore 76. Prolific plants; good flavor—harvest at 7"; excellent slicing cuke; 63 days to harvest

Straight 8. Prolific plants; 7" to 8" long fruit; excellent slicing cuke; 63 days to harvest

Suyo Longs. Burpless, sweet flavor, and crisp texture; fruit can grow to 15" long; great for slicing and pickling; 61 days to harvest

Melons (*Cucumis melo* and *Citrullus lanatus*)

You did realize that you could grow melons (even watermelons) vertically, didn't you? The varieties may not be the size that tips the scales in a contest, but they will be just as mouthwatering. The melon category includes muskmelons, watermelons, honeydews, charentias, crenshaws, and more.

Believe it or not, some of the melons listed here won't need any added support if the fruit weighs in at 2 lb. or less. If a variety produces fruit that's larger, I would use a cloth or netted sling to support them until they mature. In any case, plan to train your melons up a sturdy trellis that has a strong climbing support, such as a ladder A-frame (more in Chapter 4). If the melons mature to more than a couple of pounds, tuck a sling under each fruit and secure the sling to the climbing support. Melon varieties that are between 2 lb. and 5 or 6 lb. are good candidates for growing vertically.

GOOD TO KNOW

Did you know that the melon that North Americans refer to as a "cantaloupe" isn't a true cantaloupe at all? They're actually muskmelons. Real cantaloupes are usually grown in Europe and have a hard, warty shell. The North American muskmelons have a soft shell covered in "netting."

Planting Melons

Melon seeds can be started indoors 4 weeks before the last frost date and transplanted outdoors when soils reach 60°F to 70°F regularly. But melons grow equally well if they're planted directly into the garden bed. In both cases plant the seeds 1" deep. They require full sun and well-draining soil, and they adore nutrient-rich soil—so don't hold back on the compost. These smaller melon varieties can be spaced 2½' apart into their permanent bed.

Tending Melons

All melons love warm soil and can stop producing when the weather turns mild. Many gardeners lay black plastic down on the soil and secure it to the edges of the bed with a staple gun. Then they cut holes into the plastic and plant their young melons in the holes. This technique heats the soil up nicely and helps the melons mature quickly. I recommend adding some random slices or holes into the plastic so that water reaches the roots easily and there's some ventilation.

While the plants and immature fruit are growing, avoid intermittent watering schedules. Melons enjoy regular, even watering (not waterlogging, however) and the fruit may suffer otherwise. That fabulous melon flavor shows up in the fruit during the last 2 weeks on the vine. So at that point, ease up on the watering without letting them dry out entirely. The idea here is that they will develop less flavor if they're heavily watered close to harvest time.

Harvesting Melons

I've noticed that watermelons show several signs of being ready for harvest: if it sounds hollow when thumped with your middle finger and thumb; when the rind becomes dull, as opposed to having a youthful shine; and when the little tendril on the vine by the watermelon stem is brown and shriveled. Other melons should be harvested "on full slip," which means that the fruit slips easily off of the vine when you press your thumb onto the stem base.

Best Bets: Melon Varieties

Melons are the dessert of the vegetable garden. If you've never grown a melon variety, make this the year that you do. They're sweet, juicy, and easy to grow. Summer just wouldn't be the same without them. Each variety has a unique nectar flavor all its own.

Ambrosia. Muskmelon; mature fruit is 4 lb. to 5 lb., very netted with orange-salmon colored flesh; sugar-sweet flavor; 83 to 86 days to harvest

Delicious 51. Muskmelon; mature fruit is 3 lb. to 4 lb., aromatic fruits with orange-salmon colored flesh; delicious—tastes like sorbet; 77 days to harvest

Emerald Gem. Muskmelon; mature fruit is 2 lb., with dark green skin and orange flesh; super sweet flavor; 85 days to harvest

French Orange. Charentais cross; mature fruit is 2 lb. to 2½ lb. with netted rind and deep orange flesh; super-sweet flavor and aromatic; 75 days to harvest

Golden Midget. Watermelon; mature fruit is 3 lb., rind turns golden yellow with salmon-pink flesh; sweet flavor; 70 days to harvest

Jenny Lind. Muskmelon; mature fruit is 1 lb. to 2–lb., dark green rind with light green flesh; 70 to 85 days to harvest

Lilly. Crenshaw; mature fruit is 6 lb. to 8 lb., rind is yellow when ripe and flesh is orange; sweet and spicy flavor; 80 days to harvest

Sleeping Beauty. Muskmelon; mature fruit is tiny, ½ lb. fruits, smooth rinds turn light green just before they ripen and flesh is orange; 85 days to harvest

Tigger. Muskmelon; mature fruit is 1 lb.; vibrant yellow rind with zigzag orange stripes and white flesh; mildly sweet; 85 days to harvest

White Wonder. Watermelon; mature fruit is 3 lb. to 8 lb., green-striped rind and white flesh; sweet; 80 days to harvest

Yellow Doll. Watermelon; mature fruit is 5 lb. to 8 lb., early maturing; green-striped rinds and yellow flesh; sweet; 68 to 70 days to harvest

Peas (*Pisum sativum*)

Peas are the ladies of the spring and fall garden. Like every other vegetable I can think of, they're just so much better when they're fresh from vine to plate. These legumes are easy to grow and are perfect for vertical gardening. Peas generally fall into one (or more) of three categories:

- Shelling peas, also called English or garden peas
- Snap peas
- Snow or sugar peas

Shelling peas are usually prepared by removing the peas from inside their pods before they're cooked. With snap peas, the pods are eaten whole and are sweet and tender, even when mature. Just like green beans, they also give a great snap when bent in half. Many great snap pea varieties also make a wonderful shelling pea.

GOOD TO KNOW

I've always soaked pea and bean seeds in warm water overnight before planting the next day, and it seems to me that they come up faster. Other gardeners have mentioned the same. Just remember that it's not a deal-breaker, as most gardeners don't have any problem getting peas and beans to germinate.

Snow or sugar peas are those small peas that are associated with Asian or Chinese dishes. Pods are harvested young, and even if they're left on the vine, they don't split when they're mature like shelling and snap peas do. Like beans, peas are a legume and their roots are nitrogen-fixing.

If this is the first time peas will be grown in that bed, you may want to consider purchasing seed that's been inoculated with Rhizobium bacteria. This helps the plants fix nitrogen in their roots.

Pea vines are extremely light, and you can get away with the most basic structures as supports. Light-weight netting and twine work just fine for these climbers.

Planting Peas

Peas can be started indoors, but it's much simpler to plant them in their permanent place outdoors. Plant the seeds 1" deep and 4" apart in organically rich soil. That may sound like they're closely spaced, but they'll tolerate it—especially in loamy, prepared soil. Peas do like good air circulation, but positioning their bed in an open area makes up for the cozy spacing. Full sun is ideal, but they'll tolerate light shade without a problem.

Tending Peas

Water your peas regularly until the flowers show up; after flowering, peas require a bit more water. By "regularly" I don't mean over-watering. Just keep them evenly moist because waterlogging will slow plant growth, while drought will stress the plant and leave you with a low yield.

As far as fertilizing goes, I don't fertilize my peas much; just some compost and manure tea. I tend to aim for letting them do their thing. If you'd like to fertilize, they can use it most while they're young plants, as it takes weeks for peas to begin producing their own nitrogen.

Harvesting Peas

Harvest snow or sugar peas when they're 2" to 3" long and before the pods swell. Snap peas should be harvested after their pods swell—they'll also snap like a green bean. Shelling peas (English or Garden peas) should be harvested when they're bright green and have a cylinder shape.

No matter which peas you're growing, pick all of the mature pods as soon as you see them (and they can be hard to spot) so that the plant continues to reproduce. Also, plan on eating them as soon as possible once they've been picked; their sugars will begin to be converted to starch as soon as they're off the vine.

Best Bets: Pea Varieties

I had always been lukewarm about peas until I tasted them fresh from the garden. Then again, I can say that about several vegetables. The flavor is just better (and on occasion, quite different) when taken straight from the garden to the kitchen. The challenge is to get them from the garden to the kitchen before consuming these sweet queens of the vegetable garden.

Cascadia. Snap pea; high yield; pods are 3½" when ripe; sweet and juicy pods; 60 days to harvest

Green Arrow. Shelling pea; reliable and prolific; pods are 4" to 5" when ripe; good flavor; 68 days to harvest

Lincoln. Shelling pea; pods are 4" to 5" when ripe; excellent flavor; 67 days to harvest

Mammoth Melting Sugar. Snow pea; pods are 5" when ripe; sweet flavor; needs cool weather; 70 days to harvest

Oregon Giant. Snow pea; high yield; pods are 5" when ripe; sweet flavor; 70 days to harvest

Sugar Ann. Snap pea; early maturing, 2' vines; pods are 2½" to 3"when ripe; crisp and flavorful peas; 56 days to harvest

Super Sugar Snap. Snap pea; 6' tall, prolific vines; pods are 3" to 4" when ripe; super sweet; 62 days to harvest

Tall Telephone. Shelling pea; heat-resistant; vines reach 6' tall; long pods; sweet flavor; 75 days to harvest

Summer Squash (*Cucurbita pepo* and *maxima*)

Most people have been introduced to zucchini (over and over and over again), but have you met the other family members? They include crookneck, pattypan, scallop, crookneck, and straightneck. Try some of the more unusual summer squash varieties and enjoy some surprising colors, textures, and flavors.

Many summer squash varieties are bush types (including many zucchini varieties), which are perfect for vertical container gardens. But if you're planting a trellis or other structure, look for the vining types.

Planting Summer Squash

Squash seeds can be started indoors 3 or 4 weeks before the last frost date or they can be planted directly into the bed. Squash aren't thrilled to have their roots tampered with, but if you handle the indoor-grown starts carefully, they'll be just fine. The soil should be 70°F if you plant them directly into their bed.

Tending Summer Squash

Squash like to be grown in direct sun and soil that's rich in organic matter. Some gardeners never bother to fertilize their squash plants, and some will add some fish emulsion or compost tea once in a while. As with many other vegetables, be careful about piling on the nitrogen; doing so will yield more leaves than fruit.

Harvesting Summer Squash

Zucchini should be harvested when the fruits are about 4" to 6" long. As you may have discovered (either through experience or hearsay), zucchini is productive to say the least. If you keep the fruit harvested while they're small, it'll pump out an impressive production. But if you let one or two fully mature on the plant, it'll shut down fairly quickly (for zucchini, that is).

Pattypan and scallop squash are harvested when they're small, too—about 2" to 3" around and no bigger than 4". Harvest crookneck when they reach about 3" to 4", and straightneck when they reach 4" to 5".

Best Bets: Summer Squash Varieties

Summer temperatures may be too hot for us to handle, but summer squashes are sun-worshipers that can take the heat. They're at their best when harvested as young, immature fruit.

Black Beauty. Zucchini; prolific plant produces green-black fruit with white flesh; harvest fruit between 4" and 7" long; 45 to 65 days to harvest

Black Forest. Vining zucchini; prolific plant; harvest dark green, cylindrical fruits at 6"; 91 to 95 days to harvest

Costata Romanesco. Zucchini; pale green fruits; harvest at 6" to 10"; 52 to 60 days to harvest

Horn of Plenty. Crookneck; prolific plant; harvest fruit at 4" to 6"; 45 days to harvest

Saffron. Straightneck; heavy yields; flavorful butter-yellow fruit; harvest at 6"; 44 days to harvest

Sunburst. Pattypan; prolific plant; flavorful, yellow fruits; 52 days to harvest

Zucchetta Rampicante-Tromboncino (a.k.a. **Tromboncino**). Vining crookneck; Italian summer squash; mild and delicious flavor; harvested at 8" to 10" but can grow much longer; 80 days to harvest

Tomatoes (*Solanum lycopersicum*)

I don't think that anyone will argue when I say that tomatoes are the darlings of the American vegetable garden. I'm okay with that because I'm as smitten as anybody else.

Planting Tomatoes

If you're planting seeds, start them indoors 6 to 8 weeks before your last frost date. Seedlings can be planted in their permanent bed as soon as all danger of frost has passed and soil temps are between 55°F and 60°F. Outside temperatures need to be about 45°F or higher.

I give tomatoes about 3' of space all to themselves. It's been adequate for prepared beds—especially when they're given support with stakes or cages. Tomatoes adore sunbathing, so plant them in an area that has full sun.

Tending Tomatoes

Tomatoes do enjoy a rich soil high in organic matter, so I give it to them. This means that other than adding compost regularly, I typically don't give them a lot of fertilizer. Get them off to a good start by adding some evenly-balanced fertilizer into the planting hole along with some compost. Some gardeners also add a handful of pelletized lime, which helps to ward off blossom end rot as well as to sweeten the soil if it's a little too acidic.

A handful of bone meal is another gardener favorite for the tomato planting hole because it also combats blossom end rot as it offers calcium to the plant. After planting, I prefer not to give them anything else until the first few blossoms appear or even some fruit. I've found that when they receive too much nitrogen early on I get tons of leaves and very little fruit.

They do enjoy a well-balanced meal every so often, so once I see production on the horizon I'll apply some organic fertilizer, such as fish emulsion, about every 2 weeks until the plant stops producing. But most of the time I forget.

DOWNER

Don't randomly water tomato plants! Keep your tomato plants watered evenly and deeply because too much fluctuation between wet and dry can also set off the dreaded blossom end rot. Be sure to water them at soil level—not overhead.

Harvesting Tomatoes

With all of the fabulous varieties out there, color alone won't tell you whether your tomatoes are ripe or not. You'll know they're ready for the table when you press your finger on the skin and there's a little give. Another clue is that the skin will have a shiny appearance instead of the dull look of younger tomatoes. Try gently twisting one of the fruits that looks ripe; it should pop off the vine easily if it's ready for harvest.

DETERMINATE VERSUS INDETERMINATE TOMATO PLANTS

When it comes to choosing tomato plants, it's so tempting to focus only on the fruit size and color. But understanding the difference between determinate and indeterminate varieties will help you decide which plants are right for you.

Seed packages are usually marked as determinate or indeterminate. Nursery seedlings will also be labeled or marked "DET" (determinate) or "IND" (indeterminate). If they aren't, the nursery staff can help.

Determinate Tomatoes

The habit of a determinate tomato variety is to grow into a bush-type plant with a fixed height, although this may not be true for all of them. Once they reach about 3' to 4', the plants flower and then set fruit. This is all done in one fell swoop, and the tomatoes are harvested all at once.

Determinate varieties are great for gardeners who prefer less staking and trellising. Although a cage is usually still necessary, most of them need very little pruning—if any at all. Many hybrid or cultivated tomato plants are determinate.

For those interesting in canning or drying tomatoes, determinate varieties are a plus because the bounty is all ripe at the same time. Because of their set height, determinates are perfect for growing in containers.

Indeterminate Tomatoes

An indeterminate plant is a vine, which means it continues to grow during the entire growing season. One could argue that "true" vines are those that can attach themselves to a support as they grow, but that's merely a technicality (you say tomato …). The fact is that they tend to grow like a beanstalk as long as they're offered support.

Indeterminate vines can reach up to 12' tall and will potentially take up a lot of space unless they're pruned. Staking or caging is a must, and they'll need continued support with ties or by trellising throughout production. That said, with good pruning, they're very controllable in a small-space garden setting.

Determinates might be more compact, but indeterminate varieties have a lot going for them in their own right. These productive vines produce a higher fruit yield per square foot compared to their bushy cousins, and they win the taste test every time. In general, their tomatoes are bigger and tastier, and they continue to produce right up until a hard frost kills them. Most heirloom tomato varieties are indeterminate plants.

Indeterminate tomato varieties are perfect for the gardener who would like to use them periodically throughout the season, such as adding them to sandwiches, salads, or as side dishes. Plant more of them if you'd like to can, as well.

Best Bets: Tomato Varieties

For your vertical garden, you'll mostly be looking for indeterminate tomato varieties. But I've added some determinate types because although they won't climb much, they do grow up and some are taller than others. So they still can be very much a part of a vertical vegetable garden.

Black Krim. Indeterminate; dark, red-purple fruit; juicy, fabulous tomato flavor; 69 to 90 days to harvest

Brandywine. Indeterminate; comes in red or pink varieties; a beefsteak tomato gardener favorite; sweet tomato flavor; 90 days to harvest

Celebrity. Determinate; very reliable plant with large, red fruit; average tomato flavor; 70 days to harvest

Dad's Sunset. Indeterminate; golden-orange fruit; zesty flavor; 75 days to harvest

Early Girl. Indeterminate; 4 oz. meaty, red fruit; early season; strong tomato flavor; 75 days to harvest

Isis Candy. Indeterminate; cherry tomatoes with marbled red-gold color; super sweet flavor; 67 days to harvest

Paul Robeson. Indeterminate; black brick color; smoky, sweet, rich flavor; 90 days to harvest

Pineapple. Indeterminate; yellow skin, red marbled flesh; sweet, fruity flavor; 75 to 95 days to harvest

Purple Cherokee. Indeterminate; dark pink-purple color; super sweet flavor; 80 days to harvest

Roma VF. Determinate; one of the best paste tomato varieties; pear-shaped fruit and high yield; 75 days to harvest

San Marzano. Indeterminate; a favorite paste tomato; red; 80 days to harvest

Stupice. Indeterminate but rarely grows above 4' tall; great salad tomato; red with gold undertones; 55 to 65 days to harvest

Sun Gold. Indeterminate; apricot-orange colored cherry tomatoes; fruity flavor; 65 days to harvest

GOOD TO KNOW

When purchasing tomato plants, you may have noticed some letters on the plant's tag next to the variety name such as "VFF" or VFN." Each letter stands for a specific disease that the tomato plant variety has been bred to resist:

V = Verticillium wilt

F and FF = Fusarium wilt

N = Nematodes

T = Tobacco mosaic virus

A = Alternaria leaf spot

For example, if a tomato plant has the letter "V" next to its name (such as "Oregon Spring V"), it tells you the plant is bred to resist Verticillium wilt, which commonly attacks (and destroys) tomato crops.

Winter Squash and Pumpkins (*Cucurbita spp.*)

Just about everyone recognizes pumpkins as the round, orange winter squash that we carve into jack-o'-lanterns for Halloween. If you take another look at the pumpkin group, you'll also find that they come in colors of orange-yellow, dark green, pale green, white, red, or gray.

I've separated winter squash and pumpkins within this section because although both are the same botanically speaking, their flavors and textures are different when it comes to culinary purposes. As far as winter squash varieties, there's a lot to explore such as acorn, butternut, buttercup, spaghetti, and hubbard.

Pumpkins have a stronger flavor and coarser flesh than the other winter squash, while winter squash varieties have a milder flavor and a finer texture. Winter squash all develop a thick, hard rind or shell (some are harder than others) that's useful for winter storage. For the vertical garden you'll be most interested in pumpkins and squash varieties that mature on the smaller side. These fruits require strong supports and slings for the growing fruit.

DIY FERTILIZATION: HAND-POLLINATION

Many a gardener has joyfully spotted a little tiny fruit behind a squash flower only to have their day ruined as the little cuke shrivels up and dies a mere few days later. If this happens to you, don't label yourself as having a black thumb! The truth is that the "baby" cucumber wasn't a cucumber at all. Rather it was the promise of a cucumber—if it had only been pollinated.

Squash (or pumpkins, cucumbers, etc) have both male and female flowers. Male flowers make their appearance first and the female flowers follow. When you see a flower with a little fruit at the base, you're looking at a female. She needs to be pollinated or that little fruit will never develop into something edible.

Something (or someone) has to get the pollen from the male flower to the female flower. Unfortunately, there aren't always enough bees or other insects around to pollinate your plants. If you want to be certain that your plants are pollinated, you've got to step up to the plate!

Hand-pollinating summer and winter squash is a simple technique that comes in handy when the bees and other pollinating insects haven't done the job themselves. Before you try your hand at this (pun intended), you'll need to identify the boys from the girls. You can spot the female flower by the small "fruit" (which is really "potential" fruit) that sits just behind the blossom. The male flower lacks any such fruit.

It's as simple as taking a male flower and gently pulling off the petals. Rub the remaining male's middle part all over the female flower's middle part. You now have instant fertilization. You can also use an artist's paint brush and gently rub the bristles all around the male flower's anthers. (Those are easy to spot because they are covered with the pollen.) Then swirl the brush all over the female flower's stigma.

Planting Winter Squash and Pumpkins

If you'd like to start seeds indoors, plant them ½" to 1" deep, 3 to 4 weeks before the last frost date. To seed them directly outdoors, plant them 3' to 4' apart when soil temps reach 70°F. They enjoy a soil rich in organic matter, and add some well-rotted manure to the area to get them off to a great start.

Tending Winter Squash and Pumpkins

Squash enjoy regular watering and are heavy feeders. That said, if you add compost and some organic fertilizer several times throughout the growing season, your plants should be happy. If you'd like to add plastic over the bed to keep the soil warm (and suppress weeds), lay the plastic down first, secure the ends, cut holes in the plastic, and then plant the winter squash into the hole.

Harvesting Winter Squash and Pumpkins

When the stems of winter squash become dry and shrivel up and you can't pierce the rind of the fruit with your thumbnail, winter squash are ready to harvest. The only exception is that pumpkin rinds remain a little soft, so don't press too hard! When you harvest, be sure to leave a couple inches of stem on the squash—they'll store longer. Winter squash are best stored in a dry area at about 50°F.

Best Bets: Winter Squash and Pumpkin Varieties

Considering the amazing variety of colors, textures, and flavors, winter squash are incredibly unheralded for the home vegetable garden. Even the most popular winter squash, pumpkins, haven't been explored thoroughly by most people.

If you want to grow something new and interesting, may I suggest planting spaghetti squash (vegetable spaghetti) and preparing it for your family one winter evening. It's a delicious and fun twist on traditional pasta!

Baby Bear. Pumpkin; high yields of 6" bright orange pumpkins; great for pumpkin pie; 105 days to harvest

Burgess. Buttercup; 8" long, 3 lb. to 5 lb., turban-shaped fruit; sweet yellow-orange flesh; 90 days to harvest

Delicata. Delicata squash (also called sweet potato squash and Bohemian squash); 6" to 8" long, cream-colored rind with green stripes; sweet potato flavor; 100 days to harvest

Heart of Gold. Acorn squash; 3" to 5" fruit, cream- and dark-colored green rind with pale orange flesh; great flavor; 95 days to harvest

Jack-Be-Little. Pumpkin; vines grow to 5' and produce 3" pumpkins, both for decoration as well as eating (baked); nice mild flavor; 90 days to harvest

Mooregold. Butternut; high yields of 7" fruit with bright orange flesh and dark orange rind; sweet; 100 days to harvest

Small Sugar. Pumpkin; 7" fruits; sweet, dry, and fine flavor and texture; 100 days to harvest

Spaghetti or **Vegetable Spaghetti.** Spaghetti squash; 9" fruits have pale yellow skin and pale flesh; "spaghetti" strands are crunchy with a delicate squash flavor; 90 days to harvest

Table Queen. Acorn squash; prolific vines; 1 lb. to 3 lb.; sweet and delicate flavor; 85 days to harvest

Waltham Butternut. Butternut; vines each produce 4 to 5 squash that are 8" × 4" long and 3 lb. to 6 lb.; nutty flavor with a dry texture; 100 days to harvest

Vertically Challenged Veggies

There's more than one way to grow vegetables vertically. In this chapter, I've got the profiles on those plants that may be vertically challenged as far as vining, but I still manage to grow them in spaces far above ground level. These are some of the easiest food crops to grow in containers.

Carrots (*Daucus carota*)

Carrots aren't hard to grow, although they're notorious for taking their sweet time to germinate. Don't give up on the slowpokes of the garden, though—they're worth the time. Orange is the word of the day when it comes to carrots, but I'm out to change all of that. When you grow your own, 1,000 delicious (and beautiful) doors open. Surprisingly, carrots come in red, white, and purple, plus orange and red combos.

Planting Carrots

Before you plant carrot seeds, mix a generous amount of compost into the soil. They prefer loose and loamy sand to give their roots the wiggle room they need.

Carrots don't enjoy their roots being disturbed, so they're usually they're seeded directly into the garden bed. Sprinkle the pin-head-tiny seeds on top of loose soil 2 weeks before the last frost date. The trick with carrot seeds is to not let the soil dry out before they've germinated, so keep the seed bed evenly moist at all times.

These seeds are the tiniest little things, and some gardeners add them to sand in salt shakers, then just sprinkle them over the garden bed. I usually just broadcast and thin them out later.

Plant them ¼" to ½" deep and spaced about 2" apart in rows that are 6" to 8" apart and cover them lightly with soil. It takes 10 to 21 days for them to germinate, which feels like forever compared to other seeds.

If you want to speed things up a bit, soak them in water for 24 hours the day before you plant them. If you're stuck with heavy (clay) soul for the season, plant ball-type carrots instead of the typical, tapered, long-rooted types.

Tending Carrots

You'll need to thin the tiny seedlings when they reach about 2" high. The first time you do this it will feel brutal. But if you want to give them the room to grow properly, you'll show them the tough love.

Seedlings that are closer than 2" together are gently pulled out or snipped off with a pair of small scissors at the soil line. Using scissors is better in my opinion because it ensures that other seedling roots aren't disturbed. Toss the thinned ones into a salad or to your chickens.

GOOD TO KNOW

A good way to remind yourself to thin carrot seedlings at the right time is to plant radish seeds in between them. Radish seeds come up very fast, which reminds the gardener to water the bed; about the time that the radishes are being harvested, the carrots will need to be thinned. So there's no competition and no wasted time or garden space.

The seedlings won't need as much water as the seeds did before germination, but they do enjoy cool roots and warm shoulders (tops of carrots and the greens). When the weather really heats up, a layer of organic mulch can be tossed over them.

Harvesting Carrots

Carrots can be harvested while they're quite young (and the most tender) or when they're fully mature, which is when they're at their sweetest. Check the days to harvest for individual varieties, but most carrots are mature 2 to 3 months after their seeds have been sown.

If your soil is loose and friable, they'll slide right out with a gentile tug at the base of their green tops. But if the soil is dry or hard, use a garden fork to loosen the soil; you don't want the tops to break off as you're harvesting. Just slide the fork deep under the carrot roots and life gently. Like most vegetables, the flavor is at its best when they go straight from garden to table.

Best Bets: Carrot Varieties

One of the best things about growing carrots is that they come in many more lengths than what you'll find in a typical grocery store. There's no worries about whether you have a 12", loamy soil bed, because you can grow short, blocky, and little gourmet carrots in just a few inches of soil!

Chantenay. Blocky and broad-shouldered; adapts well to clay soils; 70 days to harvest

Danvers 126. A versatile carrot that's great for both cooking and storing; 75 days to harvest

Imperator. Store variety carrot; needs loamy soil to form well; 75 days to harvest

Little Finger. Ball-type carrot; gourmet variety that's excellent in salads; 55 days to harvest

Nantes. Crisp and sweet; stores well; 68 days to harvest

Parisienne. Ball-type carrot; French heirloom variety with excellent flavor; 50 to 68 days to harvest

Shin Kuroda. Ball-type carrot; baby Japanese heirloom variety; 75 days to harvest

Thumbelina. Ball-type carrot; great color and flavor; 60 days to harvest

Eggplants or Aubergines (*Solanum melongena*)

Most of us are familiar with the eggplants that look like large, purple teardrops. Once you start perusing the varieties, however, you'll find that they can be as small as eggs or fingers. They can be green, lavender, pink, white, or striped. Eggplants need a hot climate.

Planting Eggplants

Start your eggplant seeds indoors 4 to 6 weeks before the last frost. Some people prefer to start them directly in the garden bed because they may not transplant well. That said, I've always started them early because I needed the jump on the days they need to mature, and they've done just fine. Be warned, these seeds need a 75°F soil in order to germinate, so this would be the time to splurge on a heat mat.

Plant the seeds ¼" deep and 1" apart. Once they're outdoors in their bed, they should be planted about 2' apart from each other. With eggplant, it's always about the heat. So before you bring them outdoors, be sure that the nighttime temperatures are staying consistently above 55°F.

Tending Eggplants

Keep your eggplants watered evenly; never let their soil dry out. You can give them a balanced fertilizer once they start flowering; every other week thereafter, give them fish emulsion.

Harvesting Eggplants

Harvest your eggplants when they're fully colored (whatever color that is for the variety) and while they're still shiny. If they turn dull, you've waited too long. Use pruners and cut them away about 1" above the eggplant. Like many other vegetables, the more you harvest, the more the plants will produce.

Best Bets: Eggplant Varieties

Depending on the variety, eggplants may need to be staked or have a small pepper cage around them.

Black Beauty. 4" to 5" in diameter; shiny, black fruit; common for the market; 80 to 85 days to harvest

Casper. 6" × 2"; snow-white fruit, great for fresh summer eating; compact plants; 70 days to harvest

Dusky Hybrid. 9" × 3.5"; Deep purple color and excellent flavor; environmentally adaptable; reliable; 74 days to harvest

Fairy Tale. 5" × 1"; marbled purple and white skin; delicious creamy flavor; compact plants; 50 days to harvest

Little Fingers. 6" × 8" long when mature (but harvest at 4" to 5"); mild and sweet flavor; productive plant; purple fruits the size of fingers; 68 days to harvest

Rosa Bianca. 4" × 6"; Creamy skin with pink-purple stripes; super sweet and mild flavor; 83 days to harvest

Lettuce (Lactuca sativa)

We're all so used to store-bought lettuce that it's hard to wrap your mind around the fact that there are lettuces out there other than iceberg and your basic red leaf. Yet there are hundreds of different lettuce varieties that vary in color, shape, size, and flavor. Yes, *flavor.*

If you've never planted your own lettuce before, I urge you to give it a go this season. It's a ridiculously easy crop to grow and lightning fast. Many lettuce varieties can go from seed to plate in a mere 5 weeks.

GOOD TO KNOW

There are four general lettuce categories:

Romaine or cos lettuce produces long, thick, and crinkly leaves. It's an upright type that'll tolerate a fair amount of heat.

Butterhead or bibb lettuce has short, rather loose-leaf heads and yellow interiors. The leaves are thin and soft, and it's fairly good at tolerating some heat. You'll probably find the widest selection of heirlooms in this lettuce category.

Crispheads, Icebergs, and Batavian lettuces are the most popular types and they'll need cool weather in order to form their heads of tightly layered leaves.

Loose-leaf lettuces never form a tight head, but grow leaves that are frilly and tender. Good loose-leaf examples are Oakleaf, ruby, and the red and green leaf varieties.

Planting Lettuce

Lettuce is a cool-temperature worshipper, so plan to plant seeds in the early spring or late summer/early fall. By the way, if you have a hoop house or cold frame, lettuce can be harvested all winter long in most areas.

I sprinkle the tiny seeds onto the soil (or soil-less mix if they're indoors) and cover them lightly with soil. The seed beds are kept evenly moist. When seedling are about 1" tall, I thin them to 6" to 8" apart.

Start your first spring crop of lettuce indoors about 4 weeks before the last frost and then get them out into the garden after that last frost date has passed. The next sowing should be done directly into the garden bed in the late spring. In the fall or early winter, plant more into a cold frame; that'll let you take advantage of the sun, but will prevent the freezing temps from killing the tender leaves.

Tending Lettuce

Some lettuce varieties such as Heatwave, Black-Seeded Simpson, Red Sails, and Green Salad Bowl do quite well in warmer weather. That said, lettuce still needs a little help during higher temperatures so they don't wilt or bolt (go to seed).

The idea is to keep the soil around the lettuce as cool as possible while keeping the sun from beating down on the lettuce. Climbing vegetable crops like pole beans or tomatoes can help you out. Just plant your lettuce on the shady side of the taller dudes.

I tend to keep the soil in my lettuce beds moist at all times. Once in a while I'll water with a compost or manure tea throughout the season. Lettuces have shallow roots, so the nutrients need to be near the surface in order to benefit the plants. Fish emulsion and extra side dressings of compost every few weeks work nicely, too.

Harvesting Lettuce

Lettuce is in a rather big hurry to come to harvest. Take advantage of it by staggering your crop so you have a steady salad harvest. I know one gardener who plants lettuce in a few large pots on her patio. She plants a couple of her pots, waits a week, and then plants another couple of pots.

For tender baby greens, you can begin harvesting lettuce leaves when they're just several inches tall. However, if you're interested in harvesting when the plant is fully mature, I would suggest following the harvest dates for each variety.

There are a couple of ways to harvest lettuce:

- Head lettuce varieties are typically harvested by lifting the entire head from the soil once it becomes mature. Take note of the days to harvest for head lettuce varieties because you'll want to harvest them while they're mature, yet young.
- With the loose-leaf (romaine-types) you can harvest the outer leaves only. What's nice about this type of harvesting is that the plant can continue to grow, and you basically make salads straight from the bed. When you see a stem forming in the middle, the leaves will begin to taste bitter as the plant is beginning to bolt.
- You could also try the cut-and-come-again method with leaf lettuces. When most of the leaves have grown to the right size for salads, cut them down to about an inch or so above the soil. The plants will regrow and you'll get one or two more harvests out of that head.

Best Bets: Lettuce Varieties

Lettuces are generally cool-weather lovers, but keep your eyes open for those that are bred for some heat tolerance, which will allow you to plant closer to the summer. Otherwise, I start my lettuces in the fall and keep them going almost all winter thanks to the use of hoop houses. I also plant them in the very early spring; because lettuce matures quickly, I'm able to harvest a few crops before the summer heat sets in.

Baby Oakleaf. Looseleaf; dwarf version of its cousin Red Oakleaf; mild flavor; medium green color; 50 days to harvest

Black-Seeded Simpson. Looseleaf; traditional leaf lettuce; light and delicate flavor; very heat *in*tolerant; 49 days to harvest

Buttercrunch. Butterhead; dark green leaves with a compact head; heat tolerant; flavorful and reliable; 55 days to harvest

Deer Tongue (Matchless). Looseleaf; heavy producer of triangular leaves; upright lettuce with a crisp texture and sweet flavor; 45 to 55 days to harvest

Little Gem. Romaine; miniature green romaine with a thick heart; considered gourmet; 33 days to harvest

Marvel of Four Seasons or **Mervielle Des Quatre Saisons.** Butterhead; produces reddish rosettes; crispy with fabulous flavor; 60 days to harvest

Mascara. Butterhead; frilly, dark red, oakleaf-shaped leaves; forms a curly rosette; mild flavor; 60 to 65 days to harvest

Red Oakleaf. Looseleaf; merlot colored with mild flavor; excellent market variety; 38 days to harvest

Rouge d'Hiver (Red Winter). Romaine; red and green leaves with dark red tips and good flavor; 62 days to harvest

Peppers (*Capsicum annuum*)

Peppers include both the sweet (a.k.a. bell) and the hot types. They're available in a variety of colors, including yellow, red, green, and brown. Flavors range anywhere from mild, hot, sweet, and every nuance in between.

Next to tomatoes, peppers are the second most popular vegetable for home gardeners. Both bell and hot peppers have the same basic growing requirements and preferences, and both are good candidates for container gardening.

Planting Peppers

Peppers are usually started indoors about 8 weeks before the last frost date. They need some seriously warm soil temps for their seed to germinate; somewhere in the neighborhood of 70°F to 75°F. So using some bottom heat from a grow pad is beneficial if you can swing it.

Plant pepper seeds ¼" deep in a soil-less mix under lights. Once you get them acclimated to the outdoors, plant them about 12" apart in their bed. Peppers like to "hold hands" (leaves slightly touching) while they are mature and fruiting.

Tending Peppers

Peppers enjoy fertile, well-drained soil in the full sun. They'll also appreciate some additional compost in the bed. Give them some well-balanced fertilizer while they're young and repeat this about halfway through the growing season. Otherwise, water them regularly and deeply. Once they set fruit, they'll need a little less water than while they were growing and producing.

Harvesting Peppers

Bell peppers are ready to harvest even when they're green, but most are completely ripe when they are red, orange, yellow—whatever color that variety is supposed to be at maturity. They're also sweeter once they've gotten past the green stage (although some are meant to be green at maturity). One of the advantages of harvesting peppers when they are green is that the plant will continue to produce fruit.

DOWNER

Always wear gloves when handling hot pepper seeds, and when harvesting and washing the fruit. They'll potentially burn your skin. Whatever you do, keep your hands away from your face and eyes!

Best Bets: Pepper Varieties

Don't be afraid to let your pepper plants get cozy by planting them a little closer together than the seed packets suggest. Peppers seem to like being grown intensely, perhaps because neighboring leaves help shade the peppers.

Anaheim. Red at maturity; medium thick fruits with a pungent flavor; 80 days to harvest

California Wonder 300. Harvest when green or red; great stuffing bells; 65 days to harvest

Purple Beauty. Matures to a deep purple; succulent and productive; 70 to 80 days to harvest

Sweet Banana. Sweet wax pepper; pepper turns from yellow to red at maturity; sweet and mild; 68 days to harvest from transplant

Sweet Chocolate. Dark chocolate brown when mature; sweet and thick skin and brick-red flesh; 60 to 85 days to harvest from transplant

Sweet Golden Baby Belle. Minibell peppers at 1" to 2"; prolific plants; bright yellow; 75 to 80 days to harvest

Thai Dragon. 3" long red fruit; great for Asian dishes; fiery hot; 85 days to harvest

Potatoes (*Solanum tuberosum*)

Potatoes are in an odd category as they're most certainly sprawling plants, but aren't vining plants in the sense that they can be trained up a trellis. You wouldn't grow them in a hanging basket or most containers. But they *can* be grown up as opposed to sprawled out.

In Chapter 5 there's a section describing how to grow potatoes in a garbage bag. The same thing can be done using hooped pieces of fencing, rolls of bamboo screening, bags, boxes, baskets, and tire stacks.

Planting Potatoes

The first thing you'll want to purchase is certified disease-free seed potatoes. Using potatoes purchased at a local grocery store is discouraged, as they're not certified. Seed catalogs offer the widest assortment of varieties for home gardeners to choose from. Look for seed potatoes at local nurseries, too.

Potato plants thrive in cool weather; fertile, well-drained soil; and full sun. The soil is best on the acidic side, which helps prevent "scabs," a fungal disease.

DOWNER

Potatoes can benefit from extra soil amendments every now and then. But avoid adding straight manure because this encourages scabs (a tuber disease that leaves lesions on the skin). Use only fully composted manure or finished compost.

They can be planted whole, but usually they're cut into chunks sometimes referred to as "sets."

Each chunk should have two to three eyes on it. Before the sets are planted, they can sit around for a day or two "chitting" in a dry place so the freshly cut areas can have a chance to dry up. Plant the cut side face-down into the soil.

Dig small holes about 6" to 10" apart from one another and then plant the sets about 4" under the soil. As the plants grow, soil, straw, or leaves should be hilled up around the growing stems. Keep them covered to avoid exposure to the sun.

Potatoes that are otherwise exposed to the sun may develop solanine (a slightly toxic alkaloid). Hilling the plants (mounding the dirt up into a little hill) until there's only a little bit of the plant still sticking out encourages tuber growth.

In cooler climates, plant potatoes in early to mid-spring. If you're in a warm climate, fall is the right time for a winter/spring harvest; and a spring planting will give you a summer crop. Late winter is the right time if you're in a warm-temperate climate.

Tending Potatoes

You'll want your potato plants to have moist soil while they're growing. Because I'm always adding finished compost to the growing plants, I don't add any additional fertilizers.

Harvesting Potatoes

The gardener should stop hilling soil, leaves, or straw over the tubers when the plants begin to blossom. Adding mulch at this time is great for retaining moisture. Potatoes can be harvested when they're young, which is when the plant begins to flower. Mature potatoes are ready to harvest when the tops of the plants die down.

The potatoes should be left underground for a couple of weeks to make sure the skins have set. They're now ready to harvest for culinary dishes or to be placed into storage.

Best Bets: Potato Varieties

The potato varieties below are all excellent choices for growing "up."

All Blue. Prolific plant; indigo-colored skin with a blue flesh; delicious and colorful; 90 to 110 days to harvest

French Fingerling. Finger-shaped and rose-colored skin with yellow flesh; doesn't need to be peeled; versatile variety; 90 to 110 days to harvest

Kennebec. Large, yellow tubers; stores well; flavorful; 80 to 100 days to harvest

Red Pontiac. Red skin with white flesh; great mashing potato; stores well; 80 to 100 days to harvest

Yukon Gold. Prolific plant; yellow flesh; stores well; good flavor; 80 to 90 days to harvest

VEGGIES AND HERBS THAT TOLERATE SHADE

Most vegetables and herbs need full sun (8 hours) in order to efficiently produce, but there are quite a few exceptions. Everyone seems to have those spaces where the sun doesn't shine for very long. Here are some vegetables that have no problem with light shade. (Not all of these plants are specifically for vertical gardening, but for the sake of being thorough, I've included them.)

Crops that need 3 to 4 hours of sun:

- Arugula
- Bok choi
- Green onions (scallions)
- Herbs such as parsley, chives, garlic chives, oregano, lemon balm, cilantro, mint, and marjoram
- Kale
- Lettuce and other salad greens such as mesclun and endive
- Mustard greens
- Spinach
- Swiss chard

Crops that need 4 to 5 hours of sun:

- Beans (bush varieties)
- Beets
- Carrots
- Peas (bush varieties)
- Potatoes
- Turnips

Crops that need 5 to 6 hours of sun:

- Broccoli
- Brussels sprouts
- Cauliflower
- Radishes

Radishes (*Raphanus sativus*)

Radishes are an instant-gratification veggie. They're easy to grow, and most of them are fast maturing. We gardeners are used to waiting 80 days for ripe tomatoes and 100 days for mouthwatering watermelons; some varieties of radishes mature in a mere 25 days (although others can take as long as 60 days).

They're also available in assorted colors, such as pink, red, white, purple, rose, and yellow. Radishes have ball, cylindrical, and carrot shapes—just the cheeriest little veggies going. All of which make them the perfect first vegetable for a child's garden.

Planting Radishes

Radishes like the cool temperatures of early spring and fall, so feel free to succession plant them during these times. Prepare the soil by adding finished compost to their bed. Radishes are best planted from seed directly into the garden bed or container.

Plant them ½" deep and about 1" apart (or as close to that as possible given that they're small seeds). Once they sprout (which seems like hours) and reach a couple of inches tall, they should be thinned to every other seedling.

Tending Radishes

Radishes should be watered often and should not be left to dry out because they could crack—plus the dry soil produces hotter flavor. Feel free to add a side dressing of compost while radishes are growing or perhaps compost tea, or some fish emulsion.

Harvesting Radishes

Take note of the date to harvest that's on the seed packets of each radish variety. You'll want to harvest them as soon as they're mature. If they're left in the ground for too much longer, they'll not only acquire a sharper flavor, but will crack, as well.

Best Bets: Radish Varieties

Try planting radish seeds in the same bed (at the same time) as your carrot seeds. Carrot seeds are the turtles of the vegetable garden. Just about the time that your radishes are ready to be harvested, your carrots' sleepy little greens are up and ready to be thinned.

Cherry Belle. 2" round with bright red skin and white flesh; sweet flavor; 24 days to harvest

Chinese Red Meat. 4" round roots that resemble a watermelon when cut in half (green and white skin with red flesh); clean, sweet flavor; 30 days to harvest

Early Scarlet Globe. 1" globe with bright red skin and white flesh; extra-early harvest; 20 to 28 days to harvest

Easter Egg II. 1½" oval radishes; skin is pink, rose, white, and scarlet with white flesh; 25 days to harvest

French Dressing. 2" long × ¾" wide cylindrical, breakfast radish; bright red with white tips and white flesh; 21 days to harvest

Spinach (*Spinacia oleracea*)

Spinach is a cool-season annual plant that's frost-tolerant and easy to grow. I should also mention that the spinach you purchase at the grocery store looks innocent enough. But the truth is that it's one of the most heavily chemical-laden (pesticides) crops today. When you grow your own, you know that you're getting clean, fresh, nutritional spinach that you can feel good about serving to your family.

Planting Spinach

You can start spinach plants indoors going into a cool season, but they germinate nicely when they're planted in situ, so you may not want to bother. Seeds should be sown in the early spring or fall ½" inch deep and 1" apart. Thin the seedlings so that they end up 4" to 6" apart.

Spinach is a good candidate for succession planting; sow seeds about every week or so until you're facing the higher temperatures of summer.

Tending Spinach

Locations with soils rich in organic matter are the right places to plant spinach. In fact, if you're working with compost in the bed, you may not have to fertilize these plants at all. If the leaves become light green, feel free to give them a fertilizer that's high in nitrogen. Otherwise skip it, because if you over-fertilize spinach, it affects the flavor.

Give spinach regular water, but try to avoid the leaves so you don't encourage mildew. Mulch the base of the plants with straw or leaves to help keep the soil moist and cool.

Harvesting Spinach

Harvest the outer leaves as needed until the plant finally bolts (flowers and goes to seed). You could also cut the entire plant at the base and harvest it all at once.

Best Bets: Spinach Varieties

Spinach is a must for fall and winter gardens as it just adores the cold temperatures. It's a quick producer of some of the tastiest and most nutritious leaves in the garden. If you don't consider yourself a spinach fan, try harvesting these leaves fresh from your garden while they're young for the best flavor. You may just end up a convert.

Baby's Leaf Hybrid. Matures early; flat, green, tender leaves—lots of leaves; sweet flavor; 30 to 40 days to harvest

Bloomsdale Long Standing. Fast-maturing heirloom; crinkled, dark, glossy leaves; heavy-yielding; 40 to 48 days to harvest

Indian Summer. Dark green, crinkled leaves (savoyed); high-yielding; great flavor; 30 to 40 days to harvest

Melody. Thick, dark green, ruffled leaves; disease-resistant; 42 days to harvest

Tyee. Dark green, semi-savoyed leaves; bolts late; vigorous grower; 45 days to harvest

Swiss Chard (*Beta vulgaris*)

One of the easiest greens to grow, Swiss chard is also one of the most ornamental veggies in the garden. Planted in your front yard, no one would be able to tell that there was food among the otherwise decorative species.

Planting Swiss Chard

Swiss chard likes well-drained, fertile soils with lots of compost added, but will do fine in a sandy soil. Full sun is its preference, but it will tolerate light shade.

Sow seeds directly into the garden bed from late spring to early summer. If you live in an area that has mild winters, it can be sown in the fall and harvested all winter long.

Seeds should be planted ½" and when the seedlings show up, they should be thinned so that plants are 1' apart.

Tending Swiss Chard

Keep the soil moist at all times without waterlogging it. Once your Swiss chard is established (growing well in the garden bed) give it a balanced fertilizer and repeat about 6 to 8 weeks later.

Harvesting Swiss Chard

Swiss chard is harvested the same way as spinach: by taking the outer leaves as you need them when they reach 1' to 1½' tall. They can also be cut off at the base if you want to harvest the entire plant.

Best Bets: Swiss Chard Varieties

I find that Swiss chard is a pretty flexible plant as far as the weather is concerned. While at its best during the cool months, it can also hold its own in the warm weather, when other leafy greens such as lettuce quickly fade away.

Bright Lights. Gorgeous with stems of yellow, red, gold, rose, and white; leaves are green and burgundy; lightly savoyed leaves; 60 days to harvest

Flamingo Pink. Neon pink stalks; bright green leaves; mild flavor; 60 days to harvest

Golden Sunrise. Brilliant orange-gold stems; glowing green leaves; 2' tall plants; 55 to 65 days to harvest

Rhubarb (Ruby Red). Red stalks and green crinkled leaves; 60 days to harvest

Vertical Fruit 12

"Vertical fruit" may sound redundant given that the most common fruits grow on trees, which by definition is grown vertically. In this chapter I'm referring to fruit that's pruned using the art form of *espalier.* This is a pruning technique that keeps the tree growing flat against a wall or fence. Geometric forms make espalier beautiful as well as functional.

The wildly meandering canes of fruits such as blackberries and raspberries can also be grown vertically with the help of a trellis that holds the canes collected and upright. Then there are strawberries, which are neither a tree nor a cane, and yet they can still be contained and grown *up.*

Espalier

The most common fruit trees used for espalier are apple, pear, peach, apricot, and nectarine; although cherries, loquat, and figs can be trained into geometric shapes as well. Espalier is a term that's used in two ways. Traditionally, espalier describes a central tree (or plant) stem that's pruned so it has horizontal branch "arms" that are trained (tied) along supportive wires. In other words, the term was (and is) used to describe a specific shape.

Today, most people also use it to describe this pruning technique (which trains fruit, ornamental trees, or shrubs) no matter which shape is used. So rather than just describing the shape, it's also being used to describe the technique, too.

Espalier is practiced as an art form for the sake of design, as well as a technique to grow fruit vertically against a wall, fence, or other support. Strong supports make the best choice when working with espaliered trees, especially when they become heavy with mature fruit. Most espaliered fruit need full sun and is therefore placed on a south- or west-facing wall.

Wires are strung horizontally (12"–18" apart) against a fence, trellis, or between two poles. Fruit branches are then trained along the guide wires. Any extra branches that aren't being trained to follow the wires are trimmed off of the plant.

GOOD TO KNOW

Creating an espalier pattern can also be done with wood as guides and posts for support instead of wire guides. This is actually one of my favorite looks for a backyard because the wood adds warmth to the view. This type of trellis resembles a split-rail fence.

While this espalier form remains the most popular, it isn't your only choice by a long shot. Palmette, step-over, fan, free-form, and the crisscross-patterned Belgian fence are all beautiful in their own right. Then there are the various cordon patterns. Cordons are trees that are trained to a single "arm" and can be pruned into a single, slanted, *U*-shape, double *U*-shape, and *V*-shape.

Your first instinct may be to begin by choosing the pattern that excites you the most. Actually, the *very* first thing you need to know is which fruit you'll be planting. No matter which tree you choose, there's no doubt that you'll have some choices; but some shapes are better suited for growing certain fruits. For example, apple and pears grow best shaped into the traditional espalier, palmette, step-over, and the cordon styles. However, the stone fruits such as peaches, nectarines, and plums work best in a fan or bush shape.

Basic espalier shapes.

Many espalier shapes require regular commitment to reach their full potential as classic and beautiful focal points in your yard. I can honestly say that it's worth it in the end, but for those who want minimal work, choose a free-form shape as opposed to any of the true geometric forms.

GOOD TO KNOW

Before you purchase your tree(s) for espalier, find out if it needs to be cross-pollinated by another tree. While there are some fruits that self-pollinate and are happy as clams living life as a single, there are those that need a companion if they're to bear fruit. For example, many apple varieties need to have a different apple variety planted nearby in order to cross-pollinate. For backyard purposes, choose a semidwarf or dwarf variety.

Apples (*Malus*)

Who doesn't love an apple tree in their garden? A traditional orchard may be out of the question, but espalier-style vertical gardening makes a micro-apple-orchard possible.

Apple trees are a good example of a fruit that usually needs to cross-pollinate in order to produce. Be sure you have the space and then choose at least two different apple trees that flower at the same time. Although there are self-pollinating apple varieties such as Granny Smith and Golden Delicious, they'll bear more fruit when they're cross-pollinated. Apples are versatile as far as espalier forms go; from single to quadruple cordons, to Belgian fence, fan, and palmetto.

Planting Apple Trees

Apple trees need full sun and prefer deep, fertile, and well-drained soils, but most will do just fine in an average soil. Like other fruits, it's best for your wallet, as well as their smooth adjustment, if they're planted bare-root. That said, if I miss bare-root season, I have no problem purchasing them in soil-filled containers such as growing plants.

If you're planning on growing them in containers, be aware that if your temperatures fall below 15°F, that you may have to pull them into a protected area for the winter to protect their roots from freezing.

Apples are known for requiring long, cold winters for their fruit production. But no worries—if you live in a warmer climate, such as California, there are special varieties that produce extremely well in areas where the winters are mild and warm.

Tending Apple Trees

Apple trees like to be watered regularly and will need more in the spring and summer months during fruit development. When they're first planted, you'll also want to be certain not to let the soil dry out. One of the best ways to keep the soil damp is to place mulch around the base of the tree.

DOWNER

When you mulch the ground around your espaliered fruit trees, think "doughnut," not "volcano." Mulch piled up around a tree trunk offers easy access for critters such as snails, earwigs, and rodents (that hide under mulch) to dine on the plant. Volcano mulching can also lead to disease.

Fertilize your newly planted espaliered fruit trees with about ¼ lb. of balanced fertilizer (such as 10-10-10) right before they flower or leaf out, which is usually March. Every year after that, add another ¼ lb. of fertilizer until you reach 3 lb. If you find that your trees have 6" to 8" of tip growth every year, fertilize minimally or not at all. (See Chapter 8 for more about fertilizer.)

Harvesting Apples

It's hard to tell when an apple is mature by color alone, considering all of the different varieties available. If you pick a "test" apple and cut it open, the seeds will be pretty dark if it's mature. If the seeds are green, they're not ready for harvesting.

Another way to tell is by picking an apple by using your entire palm as opposed to using just your fingers. If you lift it and give a slight twist, a ripe apple will separate easily from the branch. The stem will also remain with the apple.

Best Bets: Espalier Apple Varieties

For good fruit production, most apple trees need the temperature to drop below 45°F for a specific period of time. This is referred to as "chilling hours." Each variety has a specific amount of chilling time that varies from short to long, so you'll want to know what the chill hours are on a fruit tree before purchasing. However, you're usually safe in purchasing from a local nursery as they've usually done the research for you and carry trees that grow well in your area. Gardeners living in mild zones (with fewer cold months) should look for apple varieties that require low chill hours.

Here are some of my favorite varieties.

Anna. Semidwarf; requires low chill hours; sweet-tart, and crisp flavor; ripens in July

Fuji. Dwarf; super sweet and crisp; ripens in September

Goldrush. Comes in both dwarf and semidwarf; excellent tart flavor that gets sweeter with time; ripens in October

Honeycrisp. Semidwarf; my favorite variety of all time; sweet, tart, juicy, and crisp; ripens starting in September

Liberty. Semidwarf; great balance between sweet and tart; ripens in September

COLUMNAR APPLE TREES

Columnar apple trees (or Colonnade apple trees) are created by the same magic as espaliered fruit trees; they're grown and pruned as a single cordon. They're the answer to a small-space gardener's prayers. These trees are shorter and much thinner than semidwarf and dwarf apple trees (although they're usually started from one or the other).

Their branchlets and short, fruiting spurs give them a "bottle-brush" appearance. Columnars usually top out at about 8' to 10' tall and their spread only 2' to 3' wide. So they can be planted a mere 2½' to 3' apart, creating a backyard (or front yard) micro-orchard. By the way, these may be mini trees, but their fruit is anything but; you'll harvest full-sized apples! They're excellent candidates for containers, too.

Another welcome characteristic is that they produce early (often in the first year) and will continue to do so for about 20 years. Like the fancier espaliered trees, handsome columnar trees accessorize a boring good-neighbor fence and complement an edible landscape.

You may need to thin the apples a bit before they fully develop to help the tree support the weight of the maturing fruit. Also, you're going to need two columnar trees for cross-pollination. They'll only produce fruit if they're cross-pollinated by a different variety, so choose one that blooms at the same time. Plant your trees near one another, either in the ground or in containers such as half-barrels.

Best Bets: Columnar apple tree varieties:

Crimson Spire. Red fruit; sweet-tart flavor; ripens in mid-September

Emerald Spire. Green fruit with gold blush; lightly sweet flavor; ripens in early to mid-September

Golden Sentinel. Yellow fruit; sweet and juicy flavor; ripens in early October

Northpole. Red fruit; crisp and juicy flavor; ripens in early September

Scarlet Sentinel. Green-yellow fruit with a red blush; sweet flavor; ripens in early October

Ultra Spire. Red fruit with yellow blush; tart and tangy flavor; ripens in mid-September

Peaches, Apricots, and Nectarines (*Prunus*)

Peaches, apricots, and nectarines all belong to the stone fruit category, along with cherries and plums. This is the hardest group to train into espalier shapes. These sweet-ladies-of-the-summer need heavy pruning and produce fruit on year-old wood; they're best pruned into a fan shape (or something informal).

Peaches, apricots, and nectarines need about 600 to 900 hours of winter chill (45°F or below) depending on the variety, although there are low-chill varieties bred for areas such as Southern California. The best place for them as espaliered specimens is against a sunny (warm) wall.

Apricots are the hardiest of the three, and there are plenty of varieties to choose from that have specific qualities, such as cold-resistance and earlier fruiting. Apricots tend to produce fruit a year or so later than peaches and nectarines. That, too, will depend upon the variety.

Planting Peaches, Apricots, and Nectarines

You can plant any of the three trees as bare-root in late winter or early spring. They can be purchased growing in containers at any time, but it's not recommended during the peak hot days of summer. Like many fruit trees, the dwarf varieties are quite suitable for containers. (Half-barrels are perfect.)

All three enjoy full sun and well-draining soil. But the soil doesn't necessarily have to be rich or loamy. In fact, they actually prefer sandy and rocky soils if they can get it. Most apricot, peach, and nectarine varieties are self-fruitful (they don't need another variety for cross-pollination in order to set fruit), but not always. So be knowledgeable about the variety you choose.

Tending Peaches, Apricots, and Nectarines

During blossom and fruit development, water them frequently and deeply. Otherwise moderate water is adequate. Feed them with a balanced fertilizer once in the early spring, right before the buds burst open. First-year trees need about ½ lb. of fertilizer; add ½ lb. every year after that until you reach 5 lb.

These trees produce excessive fruit; often more than they can even support. If you'd like large, fully formed, and healthy fruit, you'll have to do some unpleasant fruit pruning. In the spring, when you see little apricots on your tree, pinch off some of them until the fruits are 2" to 4" apart. Do the same for peaches and nectarines, but the spacing should be 3" to 5" apart. You'll thank me later.

Harvesting Peaches, Apricots, and Nectarines

Apricots should be slightly soft and have a full, apricot color when they're ready to harvest. Peaches and nectarines are ready when all of the green has colored up and they come off the branch easily, with just a slight twist of the fruit. These fruits bruise easily, so harvest them gently.

Best Bets: Peach, Apricot, and Nectarine Varieties

Like all of the fruit trees, your very best bet is to choose a variety that is known to thrive in your area.

Arctic Rose. One of the best-flavored nectarines; extremely sweet; self-fruitful; ripens in mid- to late July

Desert Dawn. Heavy producing nectarine; very sweet flavor; self-fruitful; ripens in mid- to late May

Floragold. Genetic dwarf apricot; excellent backyard fruit tree; reliable; self-fruitful; ripens in late May to early June

Goldcot. Mid- to late-season apricot; medium to large-sized fruit; self-fruitful; ripens in late June to early July

Honey Babe. Genetic dwarf peach; sweet flavor; great for home orchard; self-fruitful, but does better with Nectar Babe as a cross-pollinator; ripens in mid-July

Nectar Babe. Genetic dwarf nectarine; great pollinator for the Honey Babe peach tree; good flavor; needs another pollinator such as Honey Babe; ripens in July

Pix Zee. Genetic dwarf peach; early to mid-season; self-fruitful; ripens in mid- to late June

Puget Gold. Natural semi-dwarf apricot; good flavor; self-fruitful; ripens in early August

Reliance. Medium to large peach; flavorful and very cold-hardy; self-fruitful; ripens in mid-August

Pears (*Pyrus*)

Pears are just as versatile as apples are as far as espalier shapes are concerned. They're strong, long-lived, and typically very productive trees. That said, they're also early bloomers and may need protection from potential spring cold snaps either against a warm wall or on a slope.

With pears, you're going to need two different varieties because they're not self-fruitful.

The typical pear-shaped pears (if you will) are the European type, *Pyrus communis.* Their soft, fine flesh makes them a wonderful "dessert" fruit. Asian pears (*Pyrus pyrifolia*) resemble apples more than they do their European cousins. They have an apple-shape, crisp texture, and milder pear flavor.

Planting Pears

Pears like to live in a sunny spot that's protected from sudden spring frosts; a sunny slope is ideal. They do best in soil that's loamy and well drained; however, they tolerate heavier soils well. Again, bare-root trees planted in the late winter or early spring are ideal, but container-grown pear trees usually work out fine any time of year. Avoid planting container-grown pear trees on an extremely hot day.

Tending Pears

Pears like regular, deep watering. The feeding rule of thumb here is about 1 lb. per 1" of trunk diameter of balanced fertilizer (such as 10-10-10) in the spring. Spread it evenly at the drip line of the tree (not near the trunk).

Harvesting Pears

Patience is a virtue when it comes to pear trees, as they can start bearing fruit anywhere from 3 to 6 years after they've been planted. When they do produce, the fruit matures from summer to fall, depending on the variety. European pears should be harvested when they're *mature*, but not fully ripe. So take them from the tree while they're still firm. Allow Asian pears to fully ripen on the tree.

Best Bets: Pear Varieties

The biggest challenge with pear trees is the fire blight. It's a destructive bacterial disease that's capable of destroying limbs and sometimes the entire tree. Your best bet is to choose varieties that have been bred for good resistance.

20th Century. Asian pear with an apple shape; crisp, sweet, and juicy; medium gold skin; its excellent flavor makes it one of the most popular pear varieties today; semi-self-fruitful but does better with a cross-pollinator; ripens in August.

Hosui. Asian pear; tart and snappy flavor; fire blight resistant; self-fruiting but produces more if cross-pollinated; ripens in September

Moonglow. European pear; soft, juicy fruit with great flavor; high fire blight resistance; ripens in early August

Potomac. European pear; medium-sized fruit with green skin and red blush; fine flavor; fire blight resistant; needs a cross-pollinating tree; ripens in midseason

Warren. European pear; medium to large, pale green with red blush fruit; excellent buttery and juicy flavor; extremely fire blight resistant; self-fruitful; ripens in August

Grapes (*Vitis*)

If you enjoy grapes, I promise that you won't need a vineyard to grow them yourself. Grapes can be trained against a wall or a fence in much the same way as espaliers. In fact, one of my favorite grape supports is a chain-link fence. It's perfect for the sturdy support of thick vines, and the plants will rapidly cover the fence, creating a fruit-producing, living wall. They do very well when they're planted in large containers, as well.

They can be purchased as a bare-root plant or as a container plant. I prefer to buy them as bare-root because they're cheaper and they'll be able to become accustomed to their environment right as they break dormancy. Fences, arbors, and other strong supports are necessary for growing this vining fruit successfully.

When choosing grape varieties, the most important thing you need to remember is to pick those that grow well in your area. A local nursery, garden center, or Cooperative Extension office can steer you in the right direction. Also, most grapes are self-fruitful (with the exception of the muscadines), so feel free to purchase a single vine should you want one as a focal point or have space limitations.

Grape varieties fall into one of four categories:

- **American (*Vitis labrusca*).** This is the most commonly grown American grape species. They're known for their role as grape juice and table grapes.
- **European (*Vitis vinifera*).** Both table and wine grape varieties are found in this species.
- **Hybrids.** Hybrid grapes are crosses between European and American species. Desirable traits in crosses are easy to grow and cold tolerance.
- **Muscadine (*Vitis rotundifolia*).** Muscadine grapes are also an American species, but they rely on other vines for cross-pollination in order to produce fruit. They're also much more cold sensitive (although a couple of varieties are on the hardier side) and require a different pruning technique than other grapes. They're known for their role in wines, as well as jams.

Grapevines are vigorous climbers and become heavier as they grow and produce fruit.
(Photo courtesy of Annie Haven)

Planting Grapes

Grapes should be planted in full sun and well-draining soil about 1' to 1½' from the trellis or support and 8' to 10' apart from each other. A large pot or container can also be used in place of a true bed. Be sure to angle them slightly toward their supports. Any top growth on the plants should be cut back to where they have only 2 or 3 buds. Plant them as deep as they were in the containers they came in.

Gently pull at the roots and straighten them so that they're spread out in the planting hole. Fill in all with organically rich soil. Grapes aren't fussy about soils, but they respond well to a basically fertile one, so adding compost to the bed makes sense.

Tending Grapes

Grapes aren't heavy feeders, but if they're in containers, some liquid seaweed or other well-balanced fertilizer at least once during the season will be well received. Their leaves are prone to fungal diseases, so water grapes at the soil line, preferably with drip irrigation.

Fruit is produced on new growth, so the vines should be pruned each year. Pruning techniques will depend upon the type of grape. Grapes don't need a lot of water, but they do need *even* watering. Don't let them dry out all the way and then drench them; it stresses out the plants and the fruit will suffer.

Harvesting Grapes

Unlike some other fruits, grapes will not continue to ripen once they are picked from the vine, so don't pick them until they are fully mature. You can test for maturity by checking the color of the grapes, the size (when grapes are mature they stop growing), or simply by tasting one.

Best Bets: Grape Varieties

Because most home gardeners are interested in growing grapes for fresh eating, I've listed only table grape varieties.

Canadice. Seedless red fruit; excellent flavor; likes cool areas

Flame Seedless. The "other" most popular variety; vigorous vines; red to deep purple fruit; sweet and tart flavor

Interlaken. Seedless yellow or green grape; delicious fruity flavor

Lakemont. Seedless white grape; mild flavor

Thompson Seedless. Probably the most recognizable green table grape anywhere; sweet, mild, and juicy fruit; likes long, hot summers and doesn't do well in cool weather

Vanessa. Seeded, red fruit; crisp and fabulous flavor

Blackberries and Raspberries (*Rubus*)

Berry brambles that have been left to their own devices are unruly and seem to take over everything. It's easy to believe that a lot of space is required to have them in the home garden. But cane berries (or bramble fruits) make great vertical crops as long as you keep them in check. Plan to give them their own bed and trellis and prune them to grow in the direction you'd like—which would be up.

Planting Raspberries and Blackberries

Brambles are usually planted as a bare-root plant in the spring, although you'd be wise to consider their bed many months before that. In the fall, prepare your cane berry bed by digging the soil up about 1' deep and as long as you'd like the bed to be.

Add manure and compost to the dug soil and let it mellow by sitting undisturbed through the winter. Cane berries like to be situated in full sun with well-draining soil that's high in organic matter. They do fine with a neutral pH but prefer slightly acidic soil.

DOWNER

Don't plant your cane berries in a place where strawberries, potatoes, tomatoes, eggplants, and other brambles have been grown before. If Verticillium wilt (and other diseases) have accompanied the previous tenants, it'll come back to haunt your canes.

In the spring, plant your cane berries 3" to 10" apart (depending on the variety) in their bed along with a berry trellis. To get the most sun exposure, plant them in a row that runs north to south.

Tending Raspberries and Blackberries

Brambles tend to be leaners as opposed to true climbers. So give them the support they need with a cane berry trellis. Some raspberries are stiff-caned and are said to not need any support, but do it anyway—they'll behave better. Raspberries grow in zones 3 through 10 and blackberries will grow in zones 5 through 9.

Berries don't like to compete with weeds, so mulch them well. Water them regularly throughout the growing season. And a berry patch well-amended with organic matter won't need supplemental fertilizing. If you feel that you need to add it, fertilize just before the new growth starts.

Fruit is produced on canes in their second year, so the first year that they're planted, they won't produce fruit. Brambles that produce berries only once a year (summer/fall) should have *only the canes that have produced fruit* cut to the ground. Leave the canes that haven't produced yet undisturbed.

Canes have different names depending on if it's in its first or second year of growth. First-year canes are called primocanes and those in their second year are called floricanes. Floricanes die after they've produced fruit. Then new canes are formed from the roots (base) if the plant and the cycle begins again.

Some brambles are *everbearing*, which means that they produce one crop in the fall and a second crop in the summer. After the fall crop, prune the tops of the canes (that produced fruit) from everbearing canes. The summer crop will then be produced from the lateral canes. Once you've harvested the summer crop, cut or mow the canes to the ground. Mulch them heavily with straw for winter protection.

Harvesting Raspberries and Blackberries

Blackberries should be harvested when they're very black and firm, yet fully ripe. (Once picked, blackberries will not ripen any further.) If they're still deep red or purplish, they're not ready. Once ripe, they'll pull away from the plant easily with just a slight tug.

GOOD TO KNOW

Sometimes it's hard to tell whether you're looking at blackberries or black raspberries. The quickest way to tell the difference between a blackberry and a black raspberry is to pick a ripe fruit. You'll know the berry is ripe because it will come easily from the vine. If it's a black raspberry, the center of the fruit will be "hollow" and you'll see the core is left on the vine. When a blackberry is plucked from the bramble the core stays inside the center of the fruit and the stem that's left behind is flat and clean.

When ripe, raspberries will also slip easily from the plant. Depending on the variety, the berries may be yellow, red, or purple. Choose firm and plump berries and give a little tug. Raspberries have a hollow center as the core is left on the plant when they're pulled off. If it takes effort to harvest the berry, it's not ready yet.

DIY RECTANGLE CANE BERRY TRELLIS

There are as many variations of cane berry trellises as there are gardeners. We always have T-posts, landscape posts, wire, and heavy nylon string around our home, so that's what I use. My cane berry trellises are no-nonsense and easy to construct so I can get to planting.

Gather your materials:

4 6' metal T-posts (fence posts)

Post driver (sometimes called a T-post slammer) or mallet

Tape measure

Galvanized wire (9-gauge)

Pliers

Wire cutters

Assemble the trellis:

The T-posts will be in each corner of your rectangular berry bed, whether it's raised or just a spot on the earth that you've amended for your plants. The length of your berry bed will depend on how many plants you have and which varieties you chose.

It can be any length you'd like, but if you're planting more than 4 to 5 plants, you may want to add more T-posts so the structure remains stable as the canes grow and press against the wire.

1. Place each T-post at the bed corners and jab it hard enough into the ground so that it stands up by itself.
2. Next, use the slammer (or mallet) to hit the top of the 4 posts in order to bury the bottom 8" to 12" into the ground.
3. Using your tape measure, calculate the perimeter of the T-post-framed rectangle by adding the lengths of all four sides of the rectangle shape created by the T-posts.
4. Cut 2 pieces of galvanized wire equal to the perimeter of your bed plus an additional 8" (so you have enough wire for wrapping the ends).
5. Starting at 1 T-post, up 12" from the ground, wrap one end of the first wire around the first T-post and then secure it to the T-post with a T-clip. Twist the ends of the clip tightly together with the pliers.
6. Continue on with the wire and go around the second T-post, securing the wire to the second T-post with a clip and using the pliers to twist the end closed.
7. Repeat step 6 for the third and fourth T-posts.
8. About 24" above that first wire, repeat the whole process again with the second wire.

Now you'll have a "boxed" berry trellis that's ready to plant!

Even if you don't have these items already in your garage; all of these materials are inexpensive.

Best Bets: Raspberries and Blackberries

Here are some cane berries that are excellent for the home garden.

Cascade Delight. Red, summer-bearing raspberry; heavy producer of large, sweet fruit; excellent fresh market and backyard berry

Cherokee. Blackberry; produces medium-sized, deep black fruit with excellent flavor; vigorous and bushy

Jewel. Everbearing black raspberry; long-lived and disease-resistant; versatile, dark, and delicious fruit

Marionberry. Blackberry; vigorous plant produces large fruit with intense scent and fabulous flavor

Thornless Evergreen. Thornless blackberry; extremely popular commercial variety; very productive plant produces medium-sized fruit with a mild flavor

Tulameen. Red raspberry; one of the most popular late-season varieties; super disease resistant; produces high yields of large berries

Strawberries (*Fragaria ananassa*)

Strawberries are the queen-of-the-small-space berries. They may not climb on their own, but there's no better crop for growing in vertical containers. Strawberries that are grown specifically for food crops fall into one of three categories:

- **June-bearing:** In late spring or early summer, this group's berries all ripen at the same time and are harvested as one big crop.
- **Everbearing:** These strawberries double the fun by giving us a harvest once in early summer and again in the fall.
- **Day-neutral:** This group will produce light amounts of fruit from summer to fall. But the largest harvest is in early summer.

Planting Strawberries

The perfect spot for strawberries is in full sun with soil that's slightly acidic, well draining, and rich with organic matter. Full sun is best for the most fruit production. That said, I plant my favorite variety (Quinaults) and they've done well in light shade.

In a strawberry bed, plants are usually planted about 12" apart for each other. But when I plant them in vertical containers with loamy soil, I place them much closer together.

Strawberries shouldn't be planted flush with the soil line because it encourages crown rot. Instead, their roots should be tucked under the soil so that the crown (where the leaves come out) is sitting slightly above the soil line.

Tending Strawberries

Strawberries are happy living out their lives in a container such as a window box, hanging basket, tub, or kitchen colander. Like everyone else, I do like those big strawberry jar planters. But I skip the terra cotta variety because they dry out much too fast for my taste. I go for the glazed or plastic types so they don't dry out so quickly. For the same reason, if you go the strawberry jar route, look for the ones with the biggest pockets you can find.

Keep your plants watered just enough so that they never dry out. Drip irrigation is ideal. June-bearing plants like to have a light fertilizer feeding just before they begin to grow and right after the berries have been harvested. The rest like to be fed lightly throughout the season—light fertilization is the key with strawberry plants.

GOOD TO KNOW

After you've planted your strawberries and they begin to flower, pinch the flowers off. (No, I'm not kidding.) Without flowers, the plant can't be pollinated and this is good for a young plant because then it focuses all of its energy on building a strong root system instead of fruit production. Successful strawberry growers swear that this one little practice makes all the difference to the harvest in subsequent years.

In cold climates, you'll want to mulch them with straw or bring them under cover (which is fairly simple if they've been planted in a container).

Harvesting Strawberries

When the fruit has colored up to the point where your mouth is watering just looking at them, use your thumbnail to simply pinch through the stems. Consider choosing a variety of strawberry types so that you're collecting fruit from late spring all the way to early fall.

Best Bets: Strawberry Varieties

Homegrown strawberries fresh from the garden is one of the perks of summer eating, and I enjoy trying different varieties when I can. However, when it comes to strawberries, my best success comes when I choose those that grow well in my area. Ask your local nursery or garden center for suggestions on which strawberries will do well in your area.

Earliglow. June-bearing; one of the best-tasting varieties; disease resistant

Flamenco. Everbearing; prolific producer; excellent flavor; resistant to Verticillium wilt and powdery mildew

Ozark Beauty. Everbearing; super large, flavorful berries; high sugar content; leaf spot resistant

Quinault. Everbearing; just keeps pumping out delicious fruit; my favorite strawberry to grow for my area; excellent for hanging baskets

Reliance. June-bearing; produces high yields of smooth, glossy-red fruit; virus tolerant

Seascape. Day-neutral; naturally very sweet; high-yielding plant; disease tolerant

Sequoia. June-bearing (but acts everbearing); deep red and flavorful; resistant to powdery mildew

Tristar. Day-neutral; flavorful and large fruit; very sweet variety; disease resistant

Kiwis (*Actinidia*)

Kiwis (also called kiwifruit and Chinese gooseberries) are enjoying a growing popularity here in America, and for good reason. What kiwis lack in the good looks department they more than make up for in their sweet, tropical-tasting flesh. Kiwifruit's flavor is hard to describe, but to say it's reminiscent of a combination of other fruits such as strawberries/pineapple/melon blend is an accurate one.

Most kiwi plants are either male of female—meaning they'll bloom with either male or female flowers (dioecious). So be sure to acquire both sexes for your garden. Also, one male plant can easily pollinate six to eight females, so don't worry about keeping the numbers even. Fair warning: female plants may take 5 years to produce that first crop. But if you're a kiwi fan, it'll be worth the wait.

Typically, it's the egg-sized, fuzzy-fruited variety that you'll find in the grocery stores. These are referred to as the New Zealand kiwi, *A. deliciosa*. They're best in zones 7 through 9, and usually the skin is peeled off before eating. Once they *do* produce, they do it in spades. Fuzzy types can produce 200 fruits on one vine, and the hardy fruits about 100.

There's another less-than-common kiwi (but we're taking notice!) that's about the size of a grape called the hardy kiwi (*A. arguta*) for zones 5 through 9. Unlike the fuzzy fruits, the skin of the hardy kiwi is eaten along with the flesh. Production is about half of that of the fuzzies. There's also a kiwi called Arctic Beauty that can take even colder climates than the hardy types.

Planting Kiwis

Kiwis are vigorous vines and need a strong support as far as trellising. Pergolas, strong arbors, or other permanent trellis systems are ideal. Don't be fooled by the size of a brand-new plant—they grow to be giants.

These vines are planted as bare-root either in winter or early spring. They prefer to be in full sun, but do fine in part shade, as well. Soil should be rich in organic matter and well draining. They like their water, but they can't stand soaking in soggy soil.

Tending Kiwis

Kiwis like regular watering and won't tolerate their roots drying out. Fertilize your kiwi vines every spring with nitrogen on the high side—fish meal is ideal. Add compost to the bed and mulch to keep moisture in the soil.

For the first-year plants: plant your kiwi at the same soil level as it was in its store container. Once they're planted, mulch the bed and water it deeply. Now, trim off everything (from a female plant) but a handful of healthy buds. By the way, all we really want from the male plants are the blossoms. So prune them each year just to encourage growth and flowers.

Harvesting Kiwis

Fuzzy kiwis mature in the fall, and you'll know they're ripe when the skin changes from a green-brown to completely brown. The fruits will also start to soften. Hardy and arctic types mature in late summer to early fall. These fruits will also begin to soften when they're ready to be harvested. Of course, you could always pop one off the vine and give it the old taste test.

Best Bets: Kiwi Varieties

While all kiwis are natural vertical fruits, the Issai variety is also suitable for growing in containers. Another fancy trick of the Issai is that it's self-fruiting, plus it tends to produce the season after it's planted!

Ananasnaya. Hardy kiwi; needs to be cross-pollinated with a male plant; easy to grow; green skin with a purple-red blush; intense, pineapple-like flavor

Golden Kiwi. Fuzzy kiwi; needs to have a male kiwi nearby; bright yellow flesh; lightly tropical, very sweet flavor

Hayward. Fuzzy kiwi; requires a male plant; bright green flesh; tangy flavor; grocery store variety; popular for the home garden, as well

Issai. Hardy kiwi; self-fruiting; 1" to 1½" hairless fruit; sweet flavor; great for containers

The Vertical Herb Garden 13

When I first fell in love with gardening, herbs were my biggest seducer. Herb gardens are a win-win any way you look at it. Their presence in the kitchen is always welcome and with the exception of some shrubs and native plants, this plant group is the easiest to grow. They're beautiful, textural, and their flowers are particularly attractive to pollinating insects. Best of all, they're forgiving. We gardeners appreciate that. All herbs ask is that we water them semi-regularly and provide them with well-draining soil. Most don't require a rich soil; in fact, many do just fine in poor soils. Herbs easily shoulder gardeners' mistakes; it takes more than skipping a watering or two to make them wilt.

Even when they do show signs of stress, they seem to perk back up when the gardener is back on track. In my opinion they're the number one crop for beginning or extremely busy gardeners. This chapter contains profiles for herbs that grow well in containers—grown vertically or not.

Best Bets: Herb Containers

One of the best things about herbs is that they can be planted in anything. *Anything.* As long as there's good drainage, herbs are as adaptable as they come. You can grow them in the traditional terra cotta pots, wooden boxes, hanging baskets, 3-tiered planters, window boxes, and at the base of climbing vegetables. (See Chapter 2 for more on some of these options.) I also enjoy planting herbs in odd containers such as a shoe bag hanger, old colander, kitchen sink, and even a boot. Of course, I add drainage holes when necessary to these unique planters.

Like vegetables, many herbs can be started by seed indoors, but most of the time they're purchased as starts from a nursery or garden center. As I mentioned in Chapter 7, for some plants it's faster to make new plants from cuttings than from seed—this is true for woody-stemmed herbs such as thyme, sage, rosemary, and lemon verbena. Although they can be started from cuttings, basil, lemon balm, cilantro, and dill are quick to grow from seed.

GOOD TO KNOW

Remember that any herbs that were being housed under shade cloth, received part sun, or started indoors will have to be hardened off before they're planted into their permanent place in the garden bed or container outdoors. See Chapter 7 for details.

Basil (*Ocimum basilicum*)

Basil is not only easy to grow but it's one of the handiest herbs to have in the kitchen. Like all fresh foods, homegrown basil has the purest flavor, and if you enjoy Italian food (and by Italian, I mean tomatoes), you're going to be hooked on home-grown basil for life.

You can bet that chef Giada De Laurentis has fresh planted basil by *her* kitchen door. Basil is used in tomato, pesto, pepper, eggplant, soup, fish, and meat dishes. It's exceptional for flavoring oils and vinegars and adds something special when tossed into salads.

Basil is a fast grower and the standard varieties reach about 18" tall when mature, although dwarf and compact varieties will be about 12" or shorter. It's a bushy, tender annual with glossy leaves and blooms with tiny white or purple flower spikes. Of course, if you're growing it for culinary use, the idea is *not* to end up seeing these flower spikes, because they signal the plant to stop leaf production—which is exactly what you're harvesting. You want the plant to produce as many leaves as possible until a hard frost comes along and forces it to stop.

Various cultivars have been bred for different subtleties in flavor, appearance, and size. Keep an eye out for gorgeous purple, fine, broad, and lettuce-leaved varieties. You may also find lemon-, cinnamon-, and anise-flavored basils, as well.

Planting Basil

Basil seeds may be planted directly into the garden bed (or per garden jargon, *in situ*) after you've passed the last frost date in your area. As a native Mediterranean herb, basil likes growing in full sun, with well-drained, fertile soil. Composted manure or other organic materials tucked in for good measure is always appreciated by basil. Try to avoid over-watering the seedlings, as basil is prone to "damping off" disease.

Seeds can also be started indoors in individual little pots with a soil-less potting mix about 4 to 5 weeks *before* the last frost date. You'll have a greater success rate if they're placed on a seedling heat mat or coils, because basil both craves heat and despises cold temperatures.

GROW BASIL FROM CUTTINGS

Start basil from cuttings by filling a long-necked bottle with room-temperature water. If you have a mature basil plant in the garden, take a 6" cutting off of it (4" cutting for a dwarf basil). Now remove the lower leaves from the stem, leaving about three sets of leaves at the top.

Place the lower half of the stem into the water letting the leaves at the top hold it in place. Keep the water level high (and fresh), and in a couple of weeks you'll have new roots on your cutting.

Once the roots are a couple of inches long, take the cutting out of the bottle of water and plant it into a 4" pot with potting mix. In a couple more weeks plant the new basil plant outdoors into the garden bed or keep it as an indoor plant. If you take a cutting or two at the end of the growing season this could be a great way to bring a basil plant indoors for winter use.

Tending Basil

Once the plants are growing by several inches, you can mulch basil (as well as any other herb) with coarse mason sand. Don't buy regular playground sand—it's too fine. Mason sand is a great weed barrier and helps regulate temperature fluctuations in the bed. The best thing about using the sand as mulch in an herb bed is that it reflects the sun and douses sun-worshipping plants with heat.

While the basil is actively growing, pinch off the plant's outer leaves to encourage a bushy habit. As much as basil enjoys well-draining soil (most herbs don't like "wet feet"), it'll do best if the soil remains evenly damp. For the longest harvest try planting seeds every 2 weeks in order to keep the bounty coming. Staggering crop plantings in order to keep harvesting over a long period of time is referred to as *succession planting*.

Harvesting Basil

As soon as the basil looks abundant, you can start harvesting leaves. Just cut several inches of stems and leaves off of the plant. Watch for any flower spikes; that's a signal the plant is about to shut down production of your delicious leaves. Pinch off those spikes the minute you see them.

Best Bets: Basil Varieties

The following varieties of basil are compact (12" or under at maturity) and well-suited for vertical containers. I took the liberty of adding a couple of varieties that can reach 18" tall because they can be pruned to remain smaller and I didn't want to leave them out.

Boxwood. One of my favorites because aside from having tiny leaves and big flavor, it forms a handsome globe; 12" tall.

Cinnamon. I can't resist this one, because who doesn't love cinnamon basil? Matures up to 18" tall.

Finissimo Verde a Palla. Strong flavor along with its round form; 12"tall.

Fino Verde (little leafed). Little leaves pack a spicy-sweet flavor; 8" to 12" tall.

Green Bouquet. Tiny-leafed and compact, with pungent leaves and a sweet flavor; 8" to 12" tall.

Italian Cameo. Large leaves and a savory, rich flavor; 6" to 12" tall.

Lime. Sweet flavor with a lime zing; 12" tall.

Marseillais Dwarf. This French, extremely aromatic basil has super flavor; 12" tall.

Minette. Small-leafed and matures to a bushy globe; good basil flavor and is the traditional variety used on pizza; 8"to 10" tall.

Mrs. Burns Lemon. Strong citrus scent and lemon flavor; 12" to 18" tall.

Purple Ruffles. Ruffled, frilly, and purple; fragrant; spicy flavor (slightly anise); 12" to 18" tall.

Spicy Globe (Greek). Strong, spicy leaves with tender stems; 12" tall.

Window Box (Greek Mini). Compact, umbrella-shaped form; strong basil flavor and aroma; 8" to 10" tall.

Chives (*Allium schoenoprasum*)

Instead of producing large bulbs, perennial chives grow in tight clumps, making them the perfect candidate for containers. Both chive leaves and their flowers may be clipped off and used for their light onion flavor in any dish that calls for onions. Aside from their culinary use, they're a nice-looking plant complete with pom-pom puffs for flowers.

Planting Chives

Chives enjoy a soil that has good organic matter and drains well. They'll do best living in full sun, but have no problems with a position in the light shade.

They can be started in containers or the garden as small plants or from seed. I find it simpler to purchase them as starts, especially if I'm going to transplant them into a pot or container because you won't need very many. But they're easy to grow from seed, too.

Sow seeds indoors a few weeks before your last frost date or start them directly outdoors in early spring. Sprinkle seeds over the soil medium and cover them with about ¼" of soil and keep them evenly moist. They pop up pretty fast—as early as 7 days later.

Tending Chives

Other than regular watering and some compost or composted manure every now and again, chives require zero coddling. If you're harvesting heavily, they'll enjoy a light monthly fertilizing but that's not a hard-and-fast rule. They don't call them "the beginner's crop" for nothing.

If you grow a clump of chives in the same container for longer than a year, in the second or third year, take them out of the container and using a knife or hand shovel, cut the clump in half. Now you have a second bunch ready for another container and a fresh start for both bunches.

Harvesting Chives

If your chives were planted as starts (small plants), start harvesting after about 60 days. If you started from seed, you'll need to give them about 90 days before you start snipping. Leaves should be harvested from the *outside* of the plant, as the new leaves will grow in at the center. I often harvest my chives using the "haircut" method. Using my scissors, I simply cut straight across, being careful to only take a third of the leaves. Of course, harvesting this way leaves no chance of flowers showing up. By the way, the more blossoms you cut off, the more will bloom.

GOOD TO KNOW

Chive flowers add color, flavor, and a little surprise to salads. Another way to use them is to break the blossom apart into individual florets. Sprinkle them on foods such as potatoes, eggs, casseroles, and cooked vegetables.

Best Bets: Chive Varieties

For the most part, you'll find both the plants as well as the seed labeled as simply *chives* as opposed to having a specific variety, but there are a couple of different types.

Fine Leaf. This type is the smallest and thinnest. They're great fresh, but you have to snip quite a bit.

Garlic Chives. These grow taller than regular chives and don't have the same tubular leaves; garlic chive leaves are flat and produce white flowers with a violet scent. Distinct garlic flavor.

Purly. This type is more productive than the fine leaf and has thicker leaves.

Staro. These produce the thickest leaves and are good for freezing and drying.

Cilantro and Coriander (*Coriandrum sativum*)

Cilantro (or Chinese Parsley) is loved by many (myself included) for its aromatic leaves that add a fresh flavor to Mexican and other meat dishes. Cilantro fans don't stop there; they'll also add it to salsas, soups, curries, and chutneys.

Cilantro pulls double duty in the kitchen because its seed heads produce the culinary spice coriander. Coriander seeds have a flavor that's nutty, sweet, and spicy—all at the same time. Aside from its use in dishes, coriander can be chewed by itself as a breath freshener.

Planting Cilantro

Cilantro is another sun worshipper and does well in areas that have full sun. That said, if you happen to live in an *intensely* hot climate, it'll appreciate being planted in some light shade.

Annual cilantro can be planted directly into the garden bed from seed as soon as the last frost date in your area has gone by. Plant seeds ½" deep and 1" to 2" apart, if thinning. I find that these seeds are large, so I usually plant them about 4" apart to begin with.

If you plant them rather willy-nilly (which I admit is faster), then the little seedlings will need to be thinned out so that the remaining plants are about 4" apart. I always use a pair of little scissors when I'm thinning seedlings so that I don't disturb the tender root of the plant next to it.

If you choose to use starts, get them while they're small. Cilantro has a long taproot and doesn't enjoy being transplanted. As I mentioned above, it typically prefers full sun, but light shade is appropriate where the sun is extreme. It'll help ward off early bolting (flowering and going to seed), too. Rich, friable soil is excellent, but cilantro doesn't like lots of nitrogen. If you'd like to fertilize cilantro, use compost or well-composted manure.

Tending Cilantro

Watering evenly encourages steady growth and helps to delay bolting. I love having cilantro in my garden, but it's a loose cannon. I've never met anything that bolts so fast. The trick is to keep your eye on the plants and the minute there's a quick rise in the temperature, watch for flower spikes. I promise you that cilantro is just waiting for the opportunity to burst into flower and set seed.

I tend to take advantage of this fresh-flavored herb during the cooler weather of early spring and fall when it's in no hurry. When you notice that cilantro is starting to flower, you'll want to pinch deep into the stalk (nipping it in the bud, as it were) to ward off the inevitable.

Flowers signal the end of the lifecycle for the plant and stops leaf production, so you have to take this bolting stuff seriously if you want some goods for the kitchen. Clearly, cilantro is the poster child for succession planting.

One way to get ahead of the cilantro-bolting curve is to plant cilantro seed (coriander) in late October or early November. The seeds won't germinate during the winter, but having the seeds already planted will give you a head start because the little baby seedling will be there faster than you can find the plant on the market in the spring. This equals a longer cilantro harvest—hopefully.

Harvesting Cilantro

When left to its own devices, cilantro will reach about 1½' tall. But when you're using leaves for the kitchen, it stays significantly shorter than this. Start harvesting when the plant reaches about 8" tall. One thing to think about is that the leaves taste their best before the plant begins to bolt. Once a flower stalk or two blooms, the leaves may not be so pleasant.

One bright spot is that cilantro self-sows readily. So if (when?) it does bolt, you're sure to have more plants next season.

Harvesting Coriander

When your cilantro finally does go to seed, it's time to celebrate the coriander bounty. Coriander is a versatile spice and is used in cookies, cakes, sausages, soups, and casseroles. Seeds can be saved both for the kitchen and for next year's plants about 2 to 3 weeks after the plant has flowered. Just before the coriander begins falling naturally off the plant (they'll be a light brown), you can begin harvesting.

Simply cut the stems holding the seed heads from the plant, secure a paper bag around the whole seed head, and hang the bag upside down in a warm, well-ventilated place so they can dry for about 2 weeks.

You could also wait until you see the seeds falling from the plants, carefully cut the stalks, and then shake the rest of the seeds from the heads. Spread them into a cardboard lid that's lined with newspaper or paper towels for a couple of weeks. After your seeds are completely dry, place them in an airtight container in a cool, dark place to plant for the following season.

Best Bets: Cilantro Varieties

Most varieties vary only slightly with the exception of Leisure and Delfino, which are said to be extremely slow-bolting varieties. All are great for vertical containers. In fact, many times when you purchase cilantro from a nursery or garden center the variety isn't labeled; it simply says *cilantro.*

Delfino. Beautiful, fern-ish foliage; slow to bolt.

Leisure. A good variety for hot weather areas; slow to bolt.

Long Standing. A nice, all-around heirloom cilantro variety; slow to bolt.

Santo. Variety with heavier than normal, celery-like leaves; slow to bolt.

Mint (*Mentha spp.*)

Mint is a perennial plant that has a ton of interesting varieties and a bazillion uses in the kitchen. Their leaves are used to flavor tea, water, and other drinks. In cooking it's used to season meat dishes, breads, vegetables, jellies, fruit, pasta, and salads. Let's not forget mint's role in toothpaste, soaps, lotions, gum, and medicines.

The square-stemmed herb is perfect for pots; in fact, I won't grow them anywhere else because they seem to have an agenda—to take over the world. Mint adores our area, and thanks to their underground runners and over-enthusiastic seeds, they do a fabulous job taking over the entire garden. There are so many varieties, scents, flavors, and great ways to use mint leaves that I think they're still worth having so I keep them trapped; their roots harnessed and bound in containers.

Planting Mint

Mint propagates in almost as many ways that there are to propagate. You can start with mint seeds, but they do take a while to germinate. Plant their seeds twice as deep as they are wide and keep them moist. It's easier to start with baby mint plants, and you'll only need one per pot unless you're container is really long (or big).

A handful of roots tucked into soil will do the trick, as will random cuttings that have been set in a water glass. If you just *think* you'd like mint; you *will* have mint. Rich soil, organic matter, and sun are what mint loves, but it'll also take some shade without any problem.

Tending Mint

I try *not* to tend mine. The thought of encouraging them frightens me because it's a fact that mint will grow on concrete. Okay, I'm kidding (sort of). It prefers regular watering habits, but random is okay, too. Mint basically just goes with the flow (pun intended).

If your mint gets lanky and bare-looking (and they often do) just whack it down by at least half if not more. You're not going to kill it. In fact, you're just egging it on.

It's a good idea to divide mint in half each spring and plant half in another container. This keeps it fresh, healthy, and producing.

Harvesting Mint

Like most herbs, pick, pinch, or gather leaves whenever you need them. The more you harvest the more leaves will show up, but the flavor is always best just before they flower. The flowers are just as edible as the leaves, but I like harvesting the flower stalks as mint bouquets for inside the house.

Best Bets: Mint Varieties

Aside from all of my warnings about mint and not allowing them to run rampant, I still make room for them in my garden every year. Mints are wonderful in all of their various scents and flavors and I certainly don't want to miss out. So I give them their own container to call home and enjoy.

Apple. Fuzzy leaves have distinct apple scent; toss it into your green and fruit salads, or tea.

Banana. This banana-scented and flavored mint has round, fuzzy leaves and isn't as invasive as some of the other varieties (but I don't trust any of them).

Chocolate. This dark green-leafed mint doesn't taste and smell like pure chocolate; think peppermint dipped in chocolate. Great for desserts or an ice-cream garnish.

Lavender. The gray-green leaves are blushed purple underneath and have a strong floral scent.

Orange. This variety is nice for salads and other dishes that need a light touch. Orange mint leaves have a mild, citrus-y flavor.

Peppermint. Peppermint is strong and more potent than spearmint. Great for summer iced tea.

Pineapple. Variegated medium green and cream leaves with a pineapple scent.

Spearmint. This is the most common mint used in culinary dishes (including Mojitos).

GOOD TO KNOW

Lemon balm (*Melissa officinalis*) looks, performs, and requires the same conditions as mint, so we tend to treat it as such. Its bright green leaves are super lemon-y and popular for cool summer beverages. Lemon balm is said to be noninvasive, but I think it depends on where it's planted. Many gardeners have found it to be just as hard to control as mint, so I always pot it up just to be on the safe side.

Parsley (*Petroselinum crispum*)

Parsley basically falls into one of two categories: Italian (flat) or French (curly). While they're both hardy biennials (they flower and go to seed every other year), they're usually grown as annuals. For the sake of being thorough, there's also a "third" parsley called Hamburg parsley (var. *tuberosum*) that's not as popular as its cousins. While its greens are certainly used in cooking, Hamburg parsley is grown for its roots, which are used as a winter vegetable much like parsnips.

Both types can be used for cooking, but the Italian, flat parsley (*P. crispum* var. *neapolitanum*) is most popular for its flavor. The French, curly parsley is often added to a dish as a garnish. In my opinion, almost every dish is enhanced by parsley in one way or another, and it finds its way into sauces, rice, vegetable dishes, stews, eggs, cheese spreads, fish dishes, and more. Why not make room for both in the garden?

Planting Parsley

Stubborn parsley seeds are notoriously slow to germinate. Back in 1883 it was said that parsley has to go to the devil and back again before it will sprout and parsley has to go to the devil nine times and often forgets to come back.

You can hurry things along by soaking the seeds for 24 hours before you plant them. You can also pour boiling water over the soil or soil-less medium to speed up germination. Ideally, the soil or soil medium will be uniformly moist at about 70°F or higher. Sprinkle the tiny seeds onto the soil and cover them with about ¼" of soil. Seedlings should be thinned to 5" to 8" apart.

Because of their lazy garden attitude, many gardeners choose to start them indoors 4 to 6 weeks before the last frost. Then they're already little baby parsleys ready to transplant into their permanent bed once the soil warms up outdoors. Otherwise, just plant them directly in the garden or container in the early spring.

Tending Parsley

Parsley prefers fertile soil that's a bit on the acidic side. It likes a sunny spot where climates are cool, but tolerates light shade in the hottest areas. It enjoys a balanced fertilizer every now and again, but once the seeds have germinated, parsley isn't a fussy plant.

Harvesting Parsley

Parsley leaves can be picked or cut from the plant whenever you'd like, but it's best to harvest them from the outside of the plant so the new leaves growing on the inside have a chance to mature.

Best Bets: Parsley Varieties

In general, parsley plants grow between 12" to 18" tall (or taller). Once again, the assumption is that you'll be harvesting the leaves on this herb, which helps keep them small. Feel free to plant whatever parsley strikes your fancy in vertical containers. Like cilantro, many times you'll find them as starts and only labeled only as *flat leafed* or *curly* (without a variety name).

Banquet. French, curly

Dark Green Italian. Italian, flat-leafed

Forest Green. French, curly

Minicurl. French, curly

Moss Curled. French, curly

Single Italian. Italian flat-leafed

Oregano (*Origanum vulgare*)

Oregano is also known as "wild marjoram," although *truly* wild oregano has little or no scent. Plants or seeds found at a nursery or garden center are sure to be varieties that can be grown for culinary use. Because oregano varieties' scents (flavor) vary, rub a leaf or two before you bring it home to see if it's what you're looking for.

Oregano is a perennial herb with dark green, oval-shaped leaves that are used both fresh and dried. There are some interesting variations in leaf size and color, too, such as Aureum, which has sunny golden leaves in the spring. Standard oregano varieties grow to 1' to 2½' tall and have a potential spread of 3'. Not to worry: there are varieties bred to stay smaller, plus they become rather stunted when grown in a container, as well.

If your palate is pleased by Italian cuisine, you'll want to plant the spicier Greek or Italian variety (*Origanum vulgare hirtum*). Its gray-green leaves are broader than its cousin's and fuzzy.

Planting Oregano

You can start oregano by seed, but many people prefer to plant them as starts so they have an idea of the flavor before they're brought home. If you'd like to start them from seed, plant them indoors about 60 days before your last frost date. These seeds germinate best when they're sprinkled lightly and barely covered with soil.

If you have a mature oregano plant, you can start more plants using stem cuttings or by division (dividing the one you have into two). In a 6" vertical pot, a single plant is all you'll need. But in a larger container, plant them 6" apart.

Tending Oregano

Oregano might be the least complicated herb in the world. It requires very little attention and happily produces many pungent, aromatic leaves with only the most basic care. It absolutely thrives in rocky soils that other plants would scoff at, but it does need good drainage, and enjoys organic matter. Plant it in full sun or part shade—oregano is just not fussy about these things.

In a container filled with loamy garden soil, oregano will flourish. Water oregano evenly while it's becoming established, but after a few weeks, it requires only moderate amounts of water. Fertilizer is not on oregano's must-do list. It truly doesn't need (nor desire) it. If you have some nice compost added—it's all good.

Harvesting Oregano

In the summer or early fall, oregano blooms with little purplish-pink or white flowers. When oregano is being grown for the kitchen, the main objective is to encourage leaf production—not flowers. Are you sensing a theme here? Keep oregano trimmed so that it isn't signaled to stop producing leaves.

For best flavor, harvest the leaves before the flowers can form. That said, if I have several in my garden I like to let one or two flower to attract bees. Start taking leaf snips as soon as the plant is 4" to 5" tall.

Best Bets: Oregano Varieties

Oregano varieties typically grow between 8" and 12" tall. Like some of the other herbs in this chapter, the fact is that you shouldn't have any trouble growing any of them in vertical container systems. There's very little flavor difference between the varieties; the difference lies mainly in the color, size, and vigor.

Aureum or Golden Marjoram. Produces bright gold leaves in the spring that turn light green in the late summer or fall.

Aureum Crispum. Similar to Aureum, but with interesting, crinkly leaves.

Compact Pink Flowered. Known by its dark green, pungent leaves and dark pink flowers.

Compactum. Has leaves that turn purple in the winter.

Country Cream. Boasts variegated leaves of green with cream edging.

Greek or Italian. Recognized as the most popular variety for those who like their oregano with full-bodied flavor.

Thumble's Variety. Changes leaf color from gold-green to a medium green in the summer.

White Anniversary. Has pretty, bright green leaves with white margins—I'm hoping it wasn't named for its unremarkable white flowers.

Rosemary (*Rosemarinus officinalis*)

Rosemary is a woody, perennial herb with evergreen, needle-type leaves. Its fragrant leaves are a cook's best friend. This handsome little plant needs only the most basic care in order to look, smell, and taste wonderful.

There are two general types of rosemary: the upright and the trailing (or creeping) varieties. Trailing rosemary might create a drape immediately, hugging the ground its entire life. Or it may begin by growing up a few inches and then arch gracefully downward, which is extremely attractive coming out of pots or containers. Growing habits will depend upon the variety you choose.

All varieties are suited to culinary use and will show up with various subtleties in scent and flavor. Here in California, rosemary winters over without a problem (we have some monstrous rosemary plants). But if you're below zone 8, you'll probably have to bring it in for winter protection.

Planting Rosemary

These shrubby-looking plants are typically purchased as baby plants or taken as cuttings from a mature specimen to start new plants. Most people don't start them from seed because they're difficult to germinate, take a long time to do so, and don't always "come true" to the parent plant. Translation: you don't know what characteristics you'll get from seeds.

They enjoy full sun, well-drained soil, and aren't fond of having their roots messed with. Transplanting is clearly unavoidable, so just handle them carefully when you're moving them from one pot to another. You'll want to plant them so that the base of the plant sits a bit higher than the soil line.

Tending Rosemary

Rosemary doesn't like wet feet (soggy soil), so be sure to use a good potting soil for good drainage. It's true that rosemary is drought tolerant; however, when it's planted in a container, it's important to keep the soil a bit damp. That sounds contradictory because I just told you that it doesn't like wet feet, but there is that happy balance—I promise it's not that hard to find.

Pots dry out quickly, and rosemary can be unforgiving once it is depleted of moisture. Of course, many plants have a difficult time when they're bone dry, but rosemary finds it hard to recuperate from this and can die quickly. All of that said, I have to say that rosemary makes itself completely at home in containers and I wouldn't be without one.

This fragrant herb doesn't have a big appetite, so adding compost here and there is the best thing as far as soil conditioning. If you feel the urge to fertilize (and many do), a light application in the spring is plenty.

Harvesting Rosemary

Cut a little rosemary branch off whenever you need it, but try not to take more than ¼ to ⅓ of each branch. If you need quite a bit, then take snips from more branches instead of longer pieces of branch. The tender, new growth has the best flavor.

Here's a plus: it's not necessary to pinch the flowers off to keep the harvest going—hurray! The only pruning you might want to do here is to cut off the main shoot at the top of the plant (terminal bud) to encourage side shoot production (more leaves).

Rosemary can easily be wintered over (even if it's brought in to do so). In order to refresh the growth, use a pair of scissors in the early spring and prune it a bit. Cut off any spent flower branches and old wood, as new leaves don't grow back on old wood. The truth is that pruning rosemary is done almost solely for aesthetics (removing old wood, etc.), and while it does encourage new growth, it isn't necessary, especially in containers.

GOOD TO KNOW

Root pruning may be necessary if your rosemary plant is outgrowing its container and you'd like to keep it the same container. All you do is slide the plant out of the pot, and using a garden (or just sharp) knife, slice off about 2" of the roots from the bottom and the sides. To help the plant recover from this aggressive tactic, you should also prune the top of the plant. Add some fresh soil to the container and replant your rosemary. Between pruning and the fresh soil, your rosemary will be stimulated to get growing again!

Best Bets: Rosemary Varieties

This is an herb that can grow pretty large depending on the variety. For instance, *Barbecue* can top out at 6' tall! I thought this was important to mention considering we're discussing varieties for the purpose of vertical *containers*. Both a 6" to 12" pot and something much larger (and deeper) are acceptable as rosemary containers.

Still, your best bet will be to ask someone at the nursery to point you in the direction of the varieties that stay on the small side and have great flavor. Many nurseries and garden centers have a specific display of culinary herbs, and the perfect rosemary variety will be among that group.

This isn't to say that you couldn't try whatever variety you'd like (because they're all edible), but be aware that once it starts taking over its pot, you'll need to transplant it into a larger container or root-prune it in order to keep it in the original one. It's typical for rosemary to be transplanted at least once a year if it's in a small container.

Arp. Upright habit; lemony scent, super hardy

Blue Boy. Upright habit; teeny leaves on this teeny plant, but they're tender with good flavor and fragrance

Common rosemary. Upright habit; easy variety to find, great for the kitchen

Gorizia. Upright habit; interesting large, flat leaves

Lockwood de Forest. Trailing habit; dark green leaves, great for a cascading look

Mrs. Howard's. Unique trailing habit; branches twine and curve like no other rosemary plant

Prostratus. Trailing habit; eye-catching in hanging baskets and pots

Severn Sea. Trailing habit; frost hardy

Spice Island. Upright habit; pungent flavor and extremely fragrant foliage

Tuscan Blue. Upright habit; lemony-pine flavor and scent

Sage (*Salvia officinalis*)

Sage is a shrubby perennial plant that has a rich-spicy-earthy scent and flavor. It's a must have for cooks, and it's another easy-going, drought-tolerant plant that digs container life. Their rather thick, hairy leaves add texture and fragrance to a garden setting. They can grow anywhere from 1' to 3' tall, but like the other herbs, tend to be easily controlled in containers.

Planting Sage

Sage plants are usually purchased as one-year-old starts because harvesting their leaves without doing substantial damage to the plant begins in their second year. When transplanting them from their nursery container to your vertical garden, be sure not to plant them too deeply; the crown of the plant should sit just above soil level.

If you want to start sage from seed, plant them indoors 6 to 8 weeks before the last frost date. Sprinkle the seeds over the soil medium and cover lightly. Don't harvest leaves from sage started from seed until the second year.

Tending Sage

Sage is serious about not enjoying wet feet, so a well-draining soil is a must. They're sun worshippers but will tolerate light shade, especially where summers are extremely hot. After they're established, they need only moderate watering.

They enjoy an extra helping of compost throughout the year and a 2" to 3" application of mulch will help them winter-over. Fertilizing at the beginning of spring is usually sufficient for the year. For new growth in subsequent years, trim off the woody parts because you won't get any new leaves on old wood.

Harvesting Sage

Begin harvesting leaves for the kitchen once the plant has matured with some good growth. If it's wintered over in your container, wait for the leaves to unfold during spring to begin harvesting once again.

The same flowering principle applies here; pinch off the flower heads so leaf production doesn't slow down. I've been told that sage leaf flavor is at its best just as the flowers are opening. I can't attest to this as I've always pinched the flowers off, but it would be interesting to try.

Best Bets: Sage Varieties

Sage is fabulous for enhancing the flavors of poultry, sausage, fish, eggs (think omelets), and cheese. Harvest sage frequently to encourage the plant to become full and bushy. Sage plants that are three or four years old will become woody and scraggly. This is a good time to start over by taking a cutting from the original plant and starting fresh.

Berggarten or Mountain Garden. Purplish cast to leaves when planted in full sun; rarely flowers

Common Garden Sage. Gray-green leaves have a pebbly surface

Compacta or Nana. A narrow-leaved, compact sage

Dwarf. A small plant with gray-green leaves

Icerina. A nonblooming variety with gray-green leaves with variegated gold border

Rainbow. Purple leaves with cream and rose spots

Red Sage or Purpurascens. Spotted green, purple, and indigo leaves

Tricolor. Grayish-green leaves spotted with lilac and cream colors; new growth is tinged with purple

Thyme (*Thymus vulgaris*)

This evergreen perennial is undemanding (to say the least), yet seems to be easily overlooked both in the garden and in the kitchen. It's a sweet-looking plant that's all about tiny, dark green, grayish-green, or variegated leaves and texture. Thyme thrives naturally, living in dry, poor soil and rocky places, so it has no problem missing a watering or two. This devil-may-care attitude makes it perfectly suited to containers. Being a member of the mint family, heavily scented thyme is most popular for seasoning soups, stews, and meats.

Planting Thyme

Most people purchase them as starts, but they propagate readily from stem cuttings, too. If you'd like to start from seed, plant them indoors in late winter. Seeds should be planted about ⅛" deep in a seed-starting medium.

Once outdoors (and acclimated if they're started inside), thyme enjoys growing in full sun, but may need (or appreciate) light shade in the hottest areas. Loamy soil isn't a thyme requirement, but a well-draining soil is desirable.

Tending Thyme

Like any plant that's considered "drought tolerant," light to moderate watering should begin after the plant is established in its new home. Water new thyme plants evenly for a few weeks and then space out the time between waterings.

Thyme isn't big on specific fertilizers. If it makes you feel better, you can add some compost to be sure that the soil drains well. I wasn't kidding when I said "undemanding."

During the summer, pink, lavender, or white flowers will show up on little stalks. Of course, as far as leaf production, you'll want to pinch those off; however, I always leave some because they're excellent as pollinator attractors. If it winters over in your zone, then over time (pun intended) this little herb will become woody and less desirable for culinary use.

At this point, it's a simple thing to plant it directly into your yard somewhere and let it be the pollinating insect magnet it was born to be. Simply begin with a fresh plant in your container.

Harvesting Thyme

Once the plant is established and you have 4" to 5" of growth, snip some off anytime you need some in the kitchen.

Best Bets: Thyme Varieties

Thyme varieties grow anywhere from 2" to 18" tall, the shortest being the groundcover types. Don't let the term "ground cover" fool you, these varieties are lovely and add interest to containers, too. There are some thymes that aren't sought after for culinary use, however. It's not that they're inedible, but that they're not the best types for seasoning foods. Wooly thyme is rather … well, fuzzy. Most people don't use it for cooking. Also, elfin thyme, which is adorable in containers, is pretty tiny to use in the kitchen and is more ornamental; just something to be aware of as you shop around.

Argenteus. Has silver-edged leaves and regular oregano flavor.

Aureus or Golden Creeping Thyme. Has gold-edged foliage; lemon-scented with citrus flavor.

Caraway. Can replace caraway in your recipes.

Hi Ho. A compact thyme with variegated silver leaves.

Italian Oregano Thyme. Look at the name again. Yeah, that's confusing. This thyme literally has a distinct oregano flavor.

Lemon Thyme. Has a fresh lemon scent.

Lime. Has a lime scent; lots of green foliage.

Orange Balsam. Packed with a fresh orange scent; narrow leaves.

Pennsylvania Dutch Tea Thyme. Is fast growing with lemon-scented leaves.

Glossary  

acidic material Organic materials that fall lower than 7 on the pH (acidic/alkaline) scale of 1 through 14.

activators Organic materials, usually nitrogen or microorganisms, used to jump-start the decomposition process.

adaptability The capability to adjust readily to a climate, environment, or habitat. This is usually a process that happens over many generations.

aerate To add air; in composting, it usually involves turning the pile or adding ventilation stacks.

aerobic Those processes or organisms that can only function with oxygen present.

AHS Heat Zone Map The American Horticultural Society's (AHS) heat zone map focuses on the average highs in your zone, rather than the lows. The heat map gives you the average number of days that temperatures in your zone are at 86°F and above.

amend An action that allows you to correct or improve the soil. It can refer to the nutrition available, the organic matter, or an attempt to alter the pH balance.

anaerobic Those processes or organisms that can function without oxygen present.

annuals Plants that complete their lifecycle within a year (germinates, flowers, produces seed, and dies).

bacteria Single-celled microorganisms.

bagging A technique wherein a gardener places a bag of any kind over a pollinated flower in a way designed to keep out other pollens. This prevents cross-pollination with other plants.

beneficial insects Insects that either prey on garden pests, spread pollen to flowers in the garden, or function in any other way to help plants complete the pollination process.

biennial A plant that completes its lifecycle in 2 years. Leaves are grown the first year, and fruit and seeds are produced in the second.

biodegradable Capable of being broken down by living organisms into a simpler component.

blood meal Dried blood that's sometimes used as an activator for compost piles; it is also used as a fertilizer.

bolting When a plant produces flowers and sets seed quickly; earlier than the gardener would like them to or prematurely.

broadcast The act of spreading a soil or plant amendment like fertilizer evenly across any given area. Usually this is done by hand or a handheld spreading tool.

buffer Any compound that makes the soil less sensitive to acid and alkaline fluctuations.

C:N ratio The carbon-to-nitrogen ratio in an organic substance. *See also* carbon materials.

caging The technique of placing a cage over a plant to prevent it from cross-pollinating with another plant.

carbon materials These are usually dry materials such as straw, leaves, sawdust, cornstalks, cardboard, and paper. The balance of the C:N (carbon-to-nitrogen) ratio is 30:1. When materials fall with the "C" number being higher than 30, they're considered a carbon because their chemical makeup is predominately carbon. *See also* C:N ratio.

carnivore Any animal that primarily eats meat or meat-based food.

cloche Traditionally, a bell-shaped (often glass) cover used to protect plants from frost. More common objects are used as cloches, too, such as milk jugs with the bottom removed.

cold composting The practice of composting with minimal physical labor. You pile together browns and greens and let them sit until they decompose.

cold frame Cold frames are bottomless, box-type structures that have a transparent lid or door on the top that protects plants from cold weather. It acts as a miniature greenhouse and can extend the gardening season.

companion planting This is the practice of planting one type of plant species next to or near another—the theory being that they will benefit each other. One example is using French marigolds in the vegetable garden to help ward off nematodes under the soil.

compost Organic matter that's been biologically reduced to humus. The term is used for both the process and the end product. *See also* humus.

compost sandwich A composting technique involving layering and alternating browns and greens on a site destined to be a garden bed. No turning or adding more materials is involved.

compost tea Liquid made by "brewing" compost in a cloth; the liquid is then added to the soil or sprayed on plant leaves.

cool-season vegetables Vegetables that grow all through the cool months. These crops can be grown during two seasons: spring and fall. They need temperatures to hang around 40°F to 60°F. Examples are broccoli and cabbage.

crop rotation The practice of rotating different crop families in a garden bed or piece of land in order to control pests and disease or to increase fertility.

cross-pollination The transference of pollen from the anther (where pollen is produced) of one plant to the stigma (the part of the plant that receives the pollen) of another. This term is also used to refer to situations in which two different varieties have crossed.

cuttings Vegetatively (asexually) reproducing plants by using pieces cut from another plant.

deadheading The technique of removing spent flowers in order to let the plant focus energy on producing more blooms.

decomposer Any organism that helps break down dead plant and animal cells.

determinate (tomatoes) When referring to tomato plants, determinate (bush) describes those varieties that grow to a pre-determined size and then set their fruit all at once, within a short time. They make excellent choices for gardeners interested in canning their tomatoes. *See also* indeterminate.

diatomaceous earth A nontoxic, fine powder made from fossilized shell remains of an algae. Its sharp, microscopic edges destroy many soft-bodied insect pests.

dry processing The technique used to collect seeds from plants that produce their seeds inside a pod or husk. Good examples are peas and beans.

espalier Traditionally, a fruit tree that's been trained (pruned) to grow flat against a wall or other support. This pruning technique can also be used for other shrubs and small trees.

everbearing A term used to describe plants such as fruits or berries that don't produce in a particular season, but rather bear fruit over several seasons or every season.

F1 (filial 1) hybrid This is the offspring of two varieties (parents) that are genetically different, although they're usually within the same species.

fertilize To supply nutrients to plants.

friable soil Soil with an open structure that crumbles easily when handled. *See also* tilth.

fungi (fungus) A plant that lacks chlorophyll and vascular tissue.

genetic diversity The total range of genetic differences that are displayed within the same species. This applies to all species, including humans.

germination The very beginning of plant growth, when a seed begins to sprout into a plant.

green manure A crop (such as a legume) that's grown specifically in order to be tilled back into the soil to increase organic matter and soil fertility.

greenhouse Refers to a building especially made to house plants for protection from cold weather.

growing zone *See* USDA Hardiness Zone Map.

hand-pollination A technique that gardeners and plant breeders use to help with the pollination of their plants. This technique can include using a paintbrush to transfer pollen or shaking the branch of a self-pollinating plant.

hardening off The process of getting a plant that's been grown indoors or inside a greenhouse used to living outdoors where it will receive full exposure. Hardening off helps a plant transition with the least amount of transplant shock.

hardpan soil Topsoil that's so compacted that plant roots can't penetrate the earth.

heirloom vegetables Most gardeners agree that a variety becomes an heirloom when it's 50 years old. Most varieties are considered heirlooms when they've been handed down through the generations of a family. Heirlooms are consistently of high quality and have been saved because of their superior characteristics and ease of growing.

herbicide A synthetic chemical substance used to kill weeds.

herbivore An animal that eats plants or plant-based foods.

hoop house Usually made with flexible piping (PVC) secured over a garden bed so that plastic can be wrapped over the top to protect plants from cold weather.

hot composting The practice of keeping a compost pile hot (also known as fast), by balancing the browns (carbon), greens (nitrogen), moisture, and oxygen. This makes the temperatures toward the middle of the pile hot, which helps the compost break down quickly.

humus Material that's formed after the breakdown of organic matter. It makes complex nutrients in the soil easily accessible to plants. *See also* compost.

hybrid In the simplest terms, a hybrid is a cross between two different plant varieties in hopes of achieving the best qualities of each variety.

indeterminate (tomatoes) This describes vining tomato plants that continue to grow and bear fruit until a hard frost kills them. *See also* determinate.

inoculants Microorganisms such as fungi and bacteria that can be added to a new compost pile to help begin the decomposition process.

inorganic matter Materials that do not originate from living organisms, such as plastic and metal. *See also* organic matter.

insecticides Natural and synthetic substances that are designed to kill insects.

Integrated Pest Management (IPM) A sustainable pest management system that uses biological, cultural, and physical controls.

leaf mold The dark, earthy material that results from decomposed leaves. It's just about as close to pure humus as you can get.

lime A calcium compound made from limestone that raises the alkalinity in soil.

loamy soil A soil that generally contains more humus and therefore nutrients than sandy soils, is easier to work than clay-type soils, and has better infiltration and drainage than silty soils. The sand, silt, and clay are fairly balanced in this type of soil (40-40-20 respectively). It's considered the ideal gardening soil.

macroorganisms Organisms (animals or plants) large enough to be seen with the naked eye.

microclimates Specific local atmospheric zones where the climate differs from the larger, surrounding area.

microorganisms Organisms (animals or plants) that are too small to be seen with the naked eye.

mites Pale brown or reddish-brown insects shaped like spiders. These insects are easily seen with a hand lens.

mulch A protective covering of organic or synthetic material that's placed over the bare soil around plants to prevent weeds and erosion, retain moisture, and enrich soil.

nematodes Microscopic worms (usually), either free-living or parasitic. Some nematodes are harmful and some are helpful to plants.

nitrogen materials Materials such as grass clippings, green vegetation, manure, fruits, and vegetables that add nitrogen to a compost pile; any material with a "C" number lower than 30.

NPK Stands for nitrogen, phosphorus, and potassium, the three main nutrients plants need to thrive.

open compost pile A compost pile that is not fully contained. It's vulnerable to wildlife.

open-pollination Those plants that are produced by crossing, in nature, two parents that are the same variety. Open-pollination ends up producing nonhybrid offspring that look almost identical to their parents.

organic matter Any material that originates from living organisms, including all animal and plant life, whether still living or in any stage of decomposition. *See also* inorganic matter.

over-wintering Gardener slang for keeping a plant alive through the winter, usually by using physical barriers.

pathogens Disease-producing organisms.

peat moss A material mined from ancient bogs, it's considered a nonrenewable soil-less medium or mulch for plants.

perennial A plant that continues its lifecycle for 3 or more years. It can produce flowers and seeds from the same roots year after year.

pesticide *See* insecticides.

pH (pH scale) The acidity or alkalinity of soil as measured by the pH scale, which runs from 1 through 14. The lower the number on the scale, the higher the acidity; the higher the number, the higher the alkalinity. Soil is in balance when it falls between 6.5 and 7 on the pH scale.

photosynthesis The process in which plants make their own food by using the energy from the sun to convert carbon dioxide and water into glucose (plus other sugars and starches). The waste product plants produce is oxygen.

pinching The technique of removing the ends (terminal shoots) of actively growing plants to encourage bushy growth or give growing direction.

pistil The reproductive part of a female flower, including the style, stigma, and ovary.

potting soil A mixed medium used for planting indoor and outdoor plants in containers. May be soil-less.

propagation The process of creating new plants from parts of other plants, including cuttings, seeds, leaves, bulbs, or roots.

pruning The act of removing branches and leaves (dead or living) from a plant, shrub, or tree to improve its shape, growth, or fruit production.

row cover A thin cover that's used to cover plants to provide protection from weather, pests, or disease. They're close to the ground and may be held in place with hoops or lightly placed directly on plants.

scarification The process of wearing down hard-shelled seeds to aid in germination.

self-pollination The transference of pollen from an anther to a stigma of the same flower. It can also be the transference of pollen from the anther to the stigma of another flower on the same plant.

sheet composting *See* compost sandwich.

side dressing The process of applying nutrients in the form of fertilizer or compost onto the soil that's near a plant, but not directly under it; the nutrients are applied to the soil near a plant's stem.

stamen The male reproductive part of the flower; it carries the pollen grain and has filaments and anthers.

starts Plants that are started from seed in a nursery, brought to a garden center, and then sold to the public as baby plants.

stigma The part of the pistil (on a female flower) that receives the male pollen grains during fertilization.

stolon An above-ground plant shoot that presses horizontally against the ground and produces roots at the node.

sucker A shoot that grows from the root of a plant (usually below a graft). Suckers born on grafted plants are typically undesirable because the desired growth would be the shoots coming from at or above the graft, as opposed to the rootstock.

threshing The act of breaking the seeds free from dried plant material or seed pods.

tilth Refers to the physical condition of soil that'll be used for planting. *See also* friable soil.

top dressing A soil amendment such as compost or fertilizer applied evenly over the surface of a garden bed.

topography In regard to planting, refers to local details including natural and man-made features of the land.

trellis A frame or screen (often made of latticework) that's used as a climbing support structure for vining vegetables and ornamental plants. Loosely, other supports are referenced as trellises, such as arbors or obelisks, which may not actually be trellises in the traditional sense of the word.

USDA Hardiness Zone Map The U.S. Department of Agriculture's map dividing the United States into 11 growing areas, or zones, which are based on temperature differences of 10°F. Microclimates within a zone, as well as rainfall, day length, humidity, wind, and soil types, also play a role in planting specifics. Therefore, this map is meant to be used as a general guide.

vermicompost Worm castings mixed with bedding and partially decomposed organic matter (food). Vermicompost includes worms, their cocoons, and whatever else is living in the worm bin.

vermiculite A mineral that's used both in potting soil mixes and as a medium for rooting plant cuttings. Vermiculite aids in moisture retention and soil aeration.

warm-season vegetable Vegetables that grow through the warm months. Warm-season vegetables find their sweet spot when temperatures are above 60°F. They're usually planted during the middle to late spring or the beginning of summer. Examples are peppers and tomatoes.

wet processing A technique used to remove and dry the seeds from a pulpy fruit such as tomatoes.

winnowing A technique used to separate the grain or seed from the chaff.

Resources  

You'll find the online, product, book, and plant resources that I mentioned throughout the book in this section. I've also included websites that offer great information and tips for vertical and vegetable gardening.

Online Information

Visit the following websites and blogs for additional resources and innovative ideas for your small-space garden.

Aha! Home and Garden
ahahomeandgarden.com
Another informative blog with great reviews of products for small-space gardening.

All Thing Plants
allthingsplants.com
Plant database, blogs, tips, and best of all, *forums* to help answer all those vegetable and plant questions!

A Suburban Farmer
asuburbanfarmer.com
I couldn't leave my personal website out of the list, now could I? Its primary focus is home agriculture and small-scale gardening.

Cowlick Cottage Farm
cowlickcottagefarm.com
Food growing advice, recipes for your veggie bounty, and fabulous photographs.

Dave's Garden Freeze/Frost Dates
davesgarden.com/guides/freeze-frost-dates
Dave's Garden is a great site to visit for plant information in general. This link also offers specific information on your growing zone.

Durable Gardening
durablegardening.blogspot.com
The gardening stories at this site are great and informative—but the photos are to die for!

From the Soil
fromthesoil.blogspot.com
Go for the wonderful thoughts on plants and gardening; stay for the stunning images.

Gardening With Confidence
gardeningwithconfidence.com/blog
Excellent gardening and design advice from long-time gardener and *Better Homes and Gardens* field editor Helen Yoest.

Growing a Greener World
growingagreenerworld.com
No matter the garden size, we all strive to grow it green. This is a great blog for ideas on sustainable garden projects.

International Seed Saving Institute
seedsave.org
Dedicated to seed saving and permiculture education.

JPeterson Garden Design
jpetersongardendesign.com
Some of the most creative garden ideas anywhere are on this blog. The author, Jenny Peterson, is a landscape designer by trade, and the fabulous ideas just keep coming!

Living Homegrown
livinghomegrown.com
Award-winning garden writer and communicator Theresa Loe shares her adventures in city homesteading, edible landscapes, educational gardens, gardening with kids, and preserving the harvest.

Mulch and Soil Council
mulchandsoilcouncil.org
The Mulch and Soil Council has all the answers about the mulch and soil industry.

My Earth Garden
myearthgarden.com
Michael produces an honest blog about gardening that tells it like it is—no punches pulled here. Just the truth, excellent tips, and excellent writing.

National Gardening Association
garden.org
This organization focuses on free education in education, health and wellness, environmental stewardship, community development, and home gardening.

Our Little Acre
ourlittleacre.com
Honest home gardening insights and reviews along with beautiful images.

Punk Rock Gardens
punkrockgardens.com
This community garden blog has several writers on board. They dig up stories about gardening and good food in Pennsylvania.

Small Garden Love
smallgardenlove.com
Jacky shared her hanging herb garden design with me for this book. Her website is loaded with great DIY projects.

The Casual Gardener
thecasualgardener.blogspot.com
Fun and upbeat tips on sustainable living and a greener lifestyle.

The Grumpy Gardener
grumpygardener.southernliving.com
Gardening advice that you can hang your hat on; *Southern Living*'s senior writer, Steve Bender, is the real deal.

The Personal Garden Coach
personalgardencoach.wordpress.com
Photojournalist and garden designer Christina Salwitz offers gardening and design advice, plus gorgeous plant images.

The Rainforest Garden
therainforestgarden.com
Pure gardening inspiration and Steve Asbell's stunning artwork will have you returning again and again to The Rainforest Garden.

The Veggie Lady
theveggielady.com
Great information on growing healthy, organic vegetables in your home garden.

Urban Organic Gardener.com
urbanorganicgardener.com
Mike Lieberman takes you along on his growing adventures from his fire escape garden in New York City to his balcony in Los Angeles, California.

Vegetable Gardener
vegetablegardener.com
If you're looking for answers on growing vegetables and herbs, this is the place for you. Go ahead and post your questions—there are many experts here to answer them.

Vertical Garden Institute
verticalgardeninstitute.org
Non-profit organization dedicated to the research, education, and promotion of vertical gardens.

Product Resources

I've gathered some product websites that offer excellent products and great service.

3-D Barrel Gardens
Easiest Garden
easiestgarden.com/barrel-garden
Get step-by-step directions for several recycled, vertical garden structures.

Akro-Mils Stack-A-Pots
gardensupplyinc.com
This is the home of the stackable planters. You'll find various styles including the Stack-A-Pot as a deck rail planter.

Authentic Haven Brand Soil Conditioning Tea
ahavenbrand.com
The manure tea bags offered here are especially convenient for the small-space gardener to use for giving plants the nutrition they need.

Clean Air Gardening
cleanairgardening.com
Specializes in eco-friendly gardening tools. You'll find a ton of composting supplies here, too.

EarthBox
earthbox.com
A portable, contained gardening system that works like a dream.

Garden Harvest Supply
gardenharvestsupply.com
This online company not only offers garden supplies, but they have plant starts as well.

Gardener's Supply
gardeners.com
Gardener's Supply is an employee-owned company that has a dazzling array of gardening tools and supplies.

Garden's Alive!
gardensalive.com
Peruse environmentally responsible garden products at this site.

Good Compost
goodcompost.com
These guys have a nice selection of composters and worm bins to choose from. (Plus they're nice folks!)

Greenland Gardener
greenlandgardener.com
Excellent raised garden bed kits—plus they have some helpful gardening articles on the site.

Hanging Baskets
hangingbaskets.com/planting_info.html
This site offers a new twist on hanging baskets.

The Little Acre
myeasygrowin.com
Greatest little pop-up garden bed (bag) ever. It's durable and affordable—two of my favorite qualities.

Mobilegro
mobilegro.com
This is the Rolls Royce of garden carts—it's a beautiful product.

Planet Natural
planetnatural.com
Earth-friendly supplies for you to peruse; also has beneficial insects available for purchase.

Simply Arbors
simplyarbors.com
Arbors (and trellises) in metal, wood, iron, and vinyl.

Simply Trellises
simplytrellises.com
Trellises, arbors, lattice, and planters—they've got a vast selection here.

The Vertical Garden
theverticalgarden.com
Find vertical stacking growing containers here.

Woolly Pockets
woollypocket.com
This is the home of the 100-percent recycled plastic wall pockets, and the stand-alone "island" pockets.

Books

- Cancler, Carole. *The Home Preserving Bible: A Living Free Guide.* Indianapolis: Alpha Books, 2012.
- England, Angela. *Backyard Farming on an Acre (More or Less): A Living Free Guide.* Indianapolis: Alpha Books, 2012.
- Jabbour, Niki. *The Year-Round Vegetable Gardener.* North Adams, MA: Storey, 2011.
- McLaughlin, Chris. *The Complete Idiot's Guide to Small-Space Gardening.* Indianapolis: Alpha Books, 2012.
- ———. *The Complete Idiot's Guide to Heirloom Vegetables.* Indianapolis: Alpha Books, 2010.
- Nolan, Michael, and Reggie Solomon. *I Garden: Urban Style.* Cincinnati: Better Way Books, 2010.

- Smith, Edward C. *The Vegetable Gardener's Bible.* North Adams, MA: Storey, 2000.
- Taylor, Lisa, and Seattle Tilth. *Your Small Farm in the City.* New York, NY: Black Dog & Leventhal Publishers, 2011.

Plant Resources

Here's where you find some wonderful sources for seeds, vegetable starts, herbs, fruit trees, and cane berries.

Seeds

Baker Creek Heirloom Seeds
rareseeds.com
Baker Creek offers 1,400 heirloom varieties. All of the seeds they carry are open-pollinated, and all of them are non-GMO (free of genetically modified organisms). The photography in this catalog is ridiculously stunning. You'll want to get it for that reason alone.

BBB Seed
bbbseed.com
This company has open-pollinated, non-GMO (free of genetically modified organisms) seeds available for veggies, wildflowers, and herbs.

Beekman 1802
beekman1802.com
These guys have heirlooms, but they'll also keep you hanging around their site with the other organic goodies and information they have.

Botanical Interests
botanicalinterests.com
Botanical Interests carries both heritage and hybrid veggie, flower, and herb seeds. Lovely artwork in this catalog.

Bountiful Gardens
bountifulgardens.org
This site is all about heirlooms, open-pollination, and sustainable agriculture.

The Cook's Garden
cooksgarden.com
Cook's Garden has both heirloom and hybrids of some of the tastiest varieties.

D. Landreth Company Heirloom Seeds
landrethseeds.com
Started in 1784, Landreth is the oldest seed house in America.

Fedco Seeds
fedcoseeds.com
These guys cater to the Northeastern climate and have heirlooms and cultivar varieties available.

Kitchen Garden Seeds
kitchengardenseeds.com
Here's a great site with lots of seeds and growing tips.

The Natural Gardening Company
naturalgardening.com
In Sonoma, California, sits the oldest certified organic nursery in the United States. Come get your seeds.

Peaceful Valley
groworganic.com
They've got seeds; they've got supplies; and they've got a good attitude. What's not to love?

Reimer Seeds
reimerseeds.com
These guys have more than 4,500 quality non-GMO (free of genetically modified organisms) vegetable seeds, herb seeds, and flower seeds for the home garden and market growers.

Renee's Garden
reneesgarden.com
This is a fun site to peruse and is owned by Renee Shepherd, who is renowned in the plant world. When you order your seeds, they come in lovely little packages. The little watercolor portraits of each plant on the outside of the packaging make the wrapping as attractive as the plants that will grow in your garden.

Seed Savers Exchange
seedsavers.org
These guys offer oodles of open-pollinated seeds. If you become a member, you'll have access to thousands of heritage vegetable varieties.

Seeds of Change
seedsofchange.com
Seeds of Change has both heirloom and hybrid seeds—all are 100 percent organic.

Southern Exposure Seed Exchange
southernexposure.com
This seed company offers only varieties that are open-pollinated.

Sustainable Seed Company
sustainableseedco.com
A fantastic open-pollinated seed resource from California; 90 percent of their seeds are grown on the West Coast; 65 percent of them are grown on organic farms in the state itself. Plus, these guys are really awesome to deal with.

Terroir Seeds (Underwood Gardens)
underwoodgardens.com
This is a family-owned company that offers heirloom vegetables and flowers exclusively. It's been around since the very beginning of the heirloom seed movement.

Territorial Seed Company
territorialseed.com
This seed company has both open-pollinated seeds and hybrid seeds.

Fruit Trees and Cane Berries

Gurney's Seed & Nursery Co.
gurneys.com
This company has tons of seeds, fruit trees, and berries to choose from.

Jung Seed
www.jungseed.com
Terrific variety of fruit, vegetables, seeds, bulbs, and flowering plants are available here.

Raintree Nursery
raintreenursery.com
Find seeds, fruits, berries, mushrooms, and unusual fruits here.

Stark Bros. Nurseries & Orchards
starkbros.com
These guys have wonderful plants, berries, and fruit trees, including multi-grafted fruit trees.

Index

Numbers

A

B

C

D

E

F

G

H

I

J–K

L

M

N

O

P–Q

R

S

U–V

W–X–Y

Z